AI Management System Certification According to the ISO/IEC 42001 Standard

The book guides the reader through the certification process of the newly released ISO/IEC Artificial Intelligence standard. It provides tools and best practices on how to put together an AI management system that is certifiable and sheds light on ethical and legal challenges business leaders struggle with to make their AI system comply with existing laws and regulations, and the ethical framework of the organization.

The book is unique because it provides implementation guidance on the new certification and conformity assessment regime required by the new ISO/IEC Standard on Artificial Intelligence (ISO/IEC 42001:2023 Artificial Intelligence Management System) published by ISO and IEC in December 2023. This is the first book that addresses this issue.

As a member of the US/ISO team who participated in the drafting of this and other standard during the last three years, the author has direct knowledge and insights that are critical to the implementation of the standard. He explains the context of how to interpret ISO clauses, gives examples and guidelines, and provides best practices that help compliance managers and senior leadership understand how to put together the AI compliance system to certify their AI system. The reader will find in the book a complete guide to the certification process of AI systems and the conformity assessment required by the standard. It also provides guidance on how to read the new EU AI Act and some of the US regulations enacted by many federal agencies, state and local government entities.

This is the first book that helps the reader create an internal auditing program that enhances the company's AI compliance framework. Generative AI has taken the world by storm, and as of the writing of this book, there is no international standard that provides guidance on how to put together a management system that helps business leaders address issues of AI governance, AI structure, AI risk, AI audit, and AI impact analysis. ISO/IEC 42001:2023 is the first international voluntary certifiable standard that provides a comprehensive and well-integrated framework for the issue of AI governance. This book provides a step-by-step process on how to implement the standard so the AI system can pass the required accreditation process.

To my family, Rayan, Mona and Leila for their support and encouragement in helping me complete this book!

AI Management System Certification According to the ISO/IEC 42001 Standard

How to Audit, Certify, and Build Responsible AI Systems

Sid Ahmed Benraouane

Routledge
Taylor & Francis Group
A PRODUCTIVITY PRESS BOOK

First published 2024
by Routledge
605 Third Avenue, New York, NY 10158

and by Routledge
4 Park Square, Milton Park, Abingdon, Oxon, OX14 4RN

Routledge is an imprint of the Taylor & Francis Group, an informa business

ISBN: 978-1-032-73397-5 (hbk)
ISBN: 978-1-032-73394-4 (pbk)
ISBN: 978-1-003-46397-9 (ebk)

DOI: 10.4324/9781003463979

Typeset in Garamond
by SPi Technologies India Pvt Ltd (Straive)

Contents

Foreword

As you open this book, the first thought that may come to you is "Why do we spend millions of dollars preparing, publishing, and distributing ISO standards for everything from electric plugs to Artificial Intelligence and back again? Aren't there more than enough books and papers published on the subjects already?" Well in asking that question, you answered the question. The truth of the matter is that there are so many different ways that something can be done, used, or misused in this international consumer market that a minimum threshold on how to use or implement a new process or a new technology is required. Standards produced by the International Organization for Standardization (ISO) help organizations agree on a set of practices that make the new technology, the new process, the new service, or the new product safe and meet the quality level the consumer aspires for. The mission of ISO is to bring people together to build a consensus around new and emerging business and management practices. It provides a neutral platform where experts from different countries come together to agree on a set of ideas and ways of doing things. ISO's slogan "Great things happen when the world agrees," truly captures ISO's identity in producing frameworks that bridge barriers raised by language and cultural differences, as well as by local regulations and national legislations.

In looking back at my experience in the 1980s with ISO standards, first with ISO 9000 family (quality management), then ISO 14000 family (environmental management) and, more recently, ISO 56000 series (innovation management), I can say that with no doubt ISO standards have had a deep impact on improving the quality of products and services around the world, and have had a significant added value to process improvement. Standards have become a teaching space where entrepreneurs and managers, especially from the small and medium enterprise world, come to learn about how to apply new ideas, how to improve processes, and, ultimately, how to embrace business Excellence. Today, ISO standards serve as a benchmark for process improvement, quality control, and innovation. ISO's work brings clarity, harmony, and agreement resulting in more comprehensive interpretations of how to use and implement a new technology or a new process.

In my experience as the chair of the American Society for Quality (ASQ) and as a member of the US delegation at ISO, heavily engaged in the drafting of ISO standards, I learned many important things about the role of a standard.

First, a standard assures that the product or the service has met the quality controls that protect consumers and clarifies requirements and technical specifications to create consistency in the way technology is rolled out. A standard builds the trustworthiness needed for the product to live, grow, and prosper. A standard helps technology overcome interoperability barriers by adjusting the technology architecture to meet different contexts. Standards are a core foundation of the World Trade Organization TBT Agreements (Technical Barriers to Trade). They create the ecosystem that facilitates and promotes global trade and commerce to help the overall market integration. And finally, a standard helps in leveling off our understanding of technology to create a common language that guides us during our future scientific inquiry.

The standard you will be reading about in this book, ISO/IEC 42001 Artificial Management System (AIMS) is a game changer. It is a game changer because it addresses a transformational technology that is moving very fast, faster than we have time to understand it. Artificial Intelligence, and its most recent applications, such as Generative AI and frontier AI models, are set to challenge some of the basic foundations of the way we work and interact with each other. As a General-Purpose Technology, AI will transform everything from education to commerce, to healthcare. It is set to transform the entire global economy by adding more efficiency and productivity output. The risk of this technology going rogue is also high. Misinformation, disinformation, bias, and privacy violations, as well as performance and reputational risks, are some of the common risks listed by experts that need appropriate safeguards and oversight. The recently approved AI Act, by the EU Parliament, and the issuance of directives by federal agencies and local governments in the United States (such as the NYC Local Law 144) are a good step forward to regulate AI. But we need more. We need to go beyond setting up legal requirements. We need governance frameworks to educate managers on how to implement this technology. We need methodologies that explain to frontline employees how to interact with this technology. We need tools and best practices to help data engineers understand the ethics of producing a responsible AI model. This is what ISO/IEC 42001 brings to the table.

The book you will be reading walks you through the different requirements of the new AI Standard ISO/IEC 42001. The goal is to help you understand these requirements, link them to your business strategy, and connect them with other management systems you may have in your organization.

In the first part of the book, the author provides the general context of AI. He defines AI and discusses some of the most recent breakthroughs in AI systems. AI is also part of the broader issue, the issue of the governance of digital transformation that has become a major challenge for organizations to manage. In this part, the author sets up the scene for his analysis of the standard and provides a framework on how to engage people, modify behavior, and create a culture that aligns with the new normal of AI transformation.

Next and in the second part, the author gets into the discussion of the different clauses of the standard. He provides guidelines on how to understand and implement better the standard. While every manager has his way of implementing requirements, I sensed in the author's approach comprehensiveness, simplicity, and straightforwardness that make the implementation an easier one. He goes beyond what the standard provides to offer many options and ways of putting together an AI management system. After all, he participated in the drafting of this standard, so his insights on what fits better average organization are relevant to a good implementation. More importantly, he backs up his explanation of different clauses with examples and research from leading organizations that have set a benchmark in implementing AI management systems.

As you read this book, I want you to reflect on how your way of understanding the standard adds value to the overall AI standard development. There are many ways for you to get involved in helping others understand this new technology to make it work for everyone. I believe that we are spending too much time worrying about winners and losers. Maybe it's time to stop trying to beat each other and focus on what we can do for mankind by working together to get results. I hope you will enjoy this book.

Dr. H. James Harrington
Past President and Past Chairman of the Board
of the American Society for Quality (ASQ)
CEO of Harrington Management Systems
Los Gatos, California, USA

Preface

Generative AI: The Promise, Risks, Challenges, and Opportunities

In November 2022, the world was introduced to generative AI with the launch of ChatGPT. By January 2023, this technology was dominating headlines and blogs. To most, OpenAI had invented generative AI overnight with ChatGPT. The core technology was available for years before ChatGPT's release. But what exactly is generative AI, and how does it differ from previous AI? What promise and perils does this new wave of artificial intelligence hold? The release of ChatGPT and other generative models represents a seismic shift for AI, showcasing tremendous potential but also risks requiring thoughtful governance. Realizing the benefits of generative AI necessitates addressing core challenges and opportunities while considering the broad impact on society's digital and socioeconomic divides.

Systems like the technology behind ChatGPT – GPT-4 (Generative Pre-trained Transformer version 4) – promise to revolutionize human-machine interaction. Their ability to generate remarkably human-like text, images, code, and more opens new possibilities for automating rote work, augmenting creativity, and personalizing content. Applications span nearly every industry, from healthcare to manufacturing to entertainment, reducing costs and driving revenue. Generative models could also help tackle societal problems by uncovering insights from data. Their expansive potential explains the palpable excitement.

However, genuine risks require diligent governance, including potential bias and toxicity from imperfect training data, misinformation that appears highly credible, job displacement from rapid automation, security vulnerabilities like spreading malware or propaganda, and lack of transparency in model decision-making. Mitigating these risks necessitates ongoing research and industry collaboration to improve model architecture, training processes, human oversight, and monitoring systems.

Meaningful challenges temper unbridled enthusiasm about generative AI, including limited reasoning and brittle responses, massive data requirements many organizations lack, scarcity of qualified generative AI experts, substantial compute resources required, and building user trust in black box model recommendations. Patience is warranted as models, talent, and best practices mature.

Prudent experimentation unlocks momentous opportunities for first movers, including cost savings from automating repetitive tasks, new revenue streams with enhanced products and services, competitive advantage from cutting-edge technology, top talent attraction, customer satisfaction gains via hyper-personalization, and faster innovation and creativity. Organizations should start building foundations through small pilots focused on augmenting employees, auditing data/skills, and crafting responsible AI principles.

Democratizing access to advanced AI has tremendous potential to empower disadvantaged communities through applications like translation and intelligent tutoring. Still, there are risks that benefits accrue disproportionately to the digitally privileged as poorer communities lack infrastructure and expertise. Widespread automation also threatens to displace jobs before at-risk populations can develop new skills, and biases in training data may further sideline marginalized voices. Proactive policies and practices like investing in digital infrastructure/literacy (especially in developing nations), training models on diverse datasets, job retraining initiatives, robust safeguards against bias and misinformation, fostering inclusive innovation, and positioning as collaborators rather than replacements are essential. With ethical implementation, generative AI could uplift people worldwide, but neglecting digital and socioeconomic divides amplifies the risks of exacerbating inequality. Stakeholders must make inclusive choices to deliver on generative AI's transformative potential, unlocking immense progress for business and society. The decisions we make today will resonate for years to come.

In addition to digital and socioeconomic considerations, the environmental impact of widespread generative AI adoption warrants examination. Training and deploying these models require massive data centers, consuming substantial energy, mostly from carbon-emitting and water-consuming sources. The computational power needed also necessitates advanced semiconductors and electronics manufacturing, which is also carbon and water-intensive. While AI efficiency improves, unchecked proliferation risks increasing technology's carbon and water footprints. Environmental stewardship requires optimizing energy and water usage, using renewable power where feasible, and weighing benefits versus sustainability costs for AI applications. With mindful model development and computing strategies, generative AI can progress ecologically responsibly. However, overlooking environmental externalities will exacerbate climate change and resource scarcity. The extent to which we align generative AI with environmental values will shape its role in building a sustainable future.

The emergence of generative AI represents a turning point laden with both profound opportunities and risks. Realizing the full potential requires inclusive collaboration to develop ethical frameworks for model governance, safety, bias mitigation, job displacement, environmental impact, and other critical factors. A nuanced, thoughtful approach can unlock immense benefits for business and society. But unchecked proliferation or prioritizing narrow interests over collective welfare will undermine public trust and stunt progress. The path forward must align generative AI with shared human values of equity, responsibility, and sustainability. If stewarded judiciously, these models can augment our capabilities, increase the accessibility of information, and provide insights into serving society more broadly. Working together proactively can shape an enlightened artificial intelligence future, benefitting communities worldwide. But we must start making the right choices today.

Understanding Generative AI

Generative AI represents a paradigm shift in artificial intelligence capabilities. Rather than analyzing existing data to produce insights, generative models create entirely new data from scratch. This enables revolutionary applications across domains like natural language, images, and audio. At its core, generative AI systems are trained on massive datasets to learn generalized knowledge about the world. This foundational model can then generate or synthesize novel, original content based on prompts and parameters provided by users. The key breakthrough is producing information that may not exist in the initial training data.

For example, natural language models like GPT-4 ingest billions of web pages and books to learn linguistic skills. This training corpus contains only some potential articles, stories, or poems. But GPT-4 can then take a short text prompt and generate coherent continuations in a remarkably human-like manner. The model exhibits creativity, humor, and topical reasoning without needing specific examples of the desired output built into the initial dataset. Similarly, image generation models like DALL-E 2 and Stable Diffusion are only trained on some possible photographs. But they learn visual concepts like objects, scenes, colors, lighting, and styles from millions of images. This allows the generating of realistic and creative pictures from text prompts that were never seen before. Want an image of a unicorn playing chess in a film noir style? Generative models can synthesize it without any such photo in the training data.

Generative AI capabilities span:

- Text generation – Models like GPT-4 can produce natural language for articles, stories, emails, reports, code, and more. Prompts can specify topics, tone, length, and other parameters.
- Image generation – Create original, photorealistic images from text prompts: control image composition, style, objects, and other attributes.
- Video generation – Emerging techniques like Google's Imagen Video generate short videos from text prompts by leveraging diffusion models.
- Audio generation – Models can generate human-like voices and music by learning from large datasets. Useful for text-to-speech, voice cloning, and more.
- 3D model generation – Generate 3D shapes and scenes from text prompts. Enables rapid 3D concept model creation for gaming, VR, and design.
- Multimodal – Models combining language, image, and audio in innovative ways. For example, text-to-image-to-video generation.
- Data augmentation – Generate synthetic training data to improve downstream AI systems. Expand datasets for better machine learning.

The applications span nearly every industry:

- Chatbots – Conversational AI agents for customer service, sales, support, and more.
- Content creation – Automate writing for marketing copy, reports, news articles, social media posts, and more.
- Drug discovery – Generate molecular structures with desired pharmaceutical properties.

- Creative design – Enhance ideation and concept generation for products and media.
- Personalization – Tailor content like marketing messages, newsfeeds, and search results for each user.
- Recommendation systems – Improve product and content recommendations.
- Game development – Generate game assets like 3D models, textures, and landscapes. Assist narrative design.

The critical shift is generative AI's ability to produce novel, high-quality artifacts rather than categorize existing data. This facilitates creative tasks, personalized experiences, and democratizing access to beneficial AI applications. However, responsible development is crucial to manage risks as generative models become more capable. Generative AI can augment human capabilities and enhance our collective future if cultivated prudently.

Generative AI leverages a technique called foundation models that has rapidly advanced capabilities in recent years. But despite impressive demos, this field remains in its early stages with much room for improvement. Foundation models first train a model on a general domain like language or images. This model learns broad skills from the massive dataset without specializing in a narrow task. For example, models like GPT-3 ingest billions of web pages to acquire general linguistic abilities. The pre-trained foundation model can then be fine-tuned on smaller datasets to create specialized AI applications. This transfer learning approach enables models to learn new tasks faster with less data than training from scratch. The foundation model provides the general knowledge to bootstrap more specific capabilities.

GPT-4 demonstrates the potential of language foundation models. After training on internet text, it can generate remarkably human-like writing on most topics with simple prompting. GPT-4 powers applications like chatbots, text auto-completion, and more. Despite lacking purpose-built training, the general linguistic base enables versatile performance.

DALL-E 2 illustrates the power of image foundation models. It can generate realistic images from text prompts following pre-training on millions of photos. The general visual knowledge allows for synthesizing novel scenes and objects without explicit training examples. DALL-E fuels applications like marketing creative and product concept design. Figure I.1 provides a timeline of the recent evolution of foundation models.

Dr. Seth Dobrin CEO Qantm AI

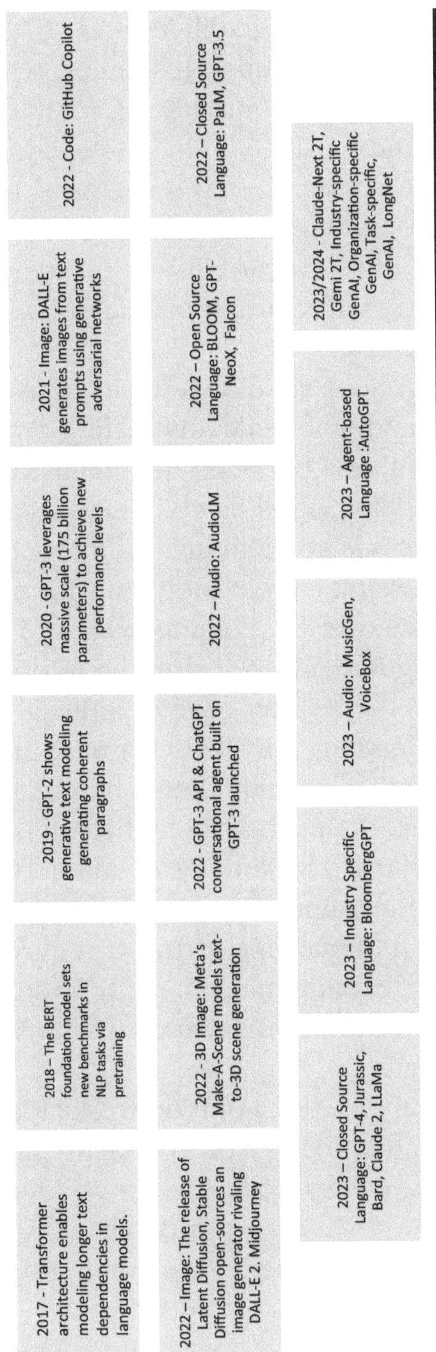

Figure I.1 A Timeline of Foundation Model Evolution.

Rapid advances in model scale, training data, and computing power have enabled the proliferation of generative AI foundation models. Still, significant limitations remain, including brittleness outside the original training distribution, limited reasoning abilities and common sense, heavy reliance on extensive data and computing resources, a propensity for generating misinformation or biased outputs, inscrutable decision-making processes, and high costs and barriers to entry. Overcoming these challenges will require dedicated research into robustness, reasoning, transparency, efficiency, and accessibility to realize the full potential of generative AI while mitigating risks. Further research is critical to address these challenges. Key priorities include improving robustness, reasoning, and transparency while expanding access.

In the near term, organizations should view generative AI as augmenting human capabilities rather than wholesale automating processes. These models excel at creative synthesis but need help with complex inference and judgment. Uncritically deferring to flawed outputs can cause genuine harm. But the rate of progress is accelerating. With thoughtful stewardship, generative foundation models may profoundly transform how humans leverage artificial intelligence over the coming decade. The seeds have been planted, but much nurturing remains to realize the full potential.

Generative AI represents the leading edge of artificial intelligence today. It builds on prior machine learning and deep learning innovations while possessing distinct capabilities. Understanding where generative models fit into the broader AI landscape helps illuminate their strengths and limitations. At a high level, generative AI leverages neural networks, a technique under the umbrella of deep learning. Deep learning emerged in the 2000s as a powerful new approach to machine learning. It introduced multilayer neural networks capable of learning hierarchical representations from data automatically. This enabled breakthroughs in computer vision, speech recognition, and other domains.

Generative modeling adapts deep neural networks for imaginative tasks like creating new content rather than just recognizing patterns in data. Architectures like generative adversarial networks (GANs) and variational autoencoders (VAEs) empower generative capabilities. Models can synthesize novel images, text, audio, and video after training on large datasets of examples.

Whereas deep learning analyzes existing data, generative AI creates new artifacts that differ from training data. This creative aspect expands the possibilities for innovating, designing, and personalizing. Generative models also incorporate mechanisms like attention that improve their ability to model relationships and context.

However, current generative AI techniques still have significant limitations compared to human cognition. They need a deeper understanding of the concepts and meaning behind the content they generate, with models manipulating symbols without fully grounding them in the real world. Their reasoning abilities could be more profound, needing help to perform complex inference or reasoning that links disparate knowledge. This reliance on pattern recognition from training data also leads to brittleness when presented with unusual inputs. The lack of common sense allows models to generate logically inconsistent or unethical outputs without human rationality and ethical constraints. Overcoming these shortcomings through hybrid approaches combining neural techniques with symbolic knowledge representation and reasoning will be essential frontiers in developing more robust and beneficial generative AI. These shortcomings contrast with earlier approaches to AI, like symbolic systems and knowledge representation. Symbolic AI aims to represent knowledge with semantic models and logical inference rules explicitly. This allows more explainable reasoning and more flexible knowledge manipulation.

Integrating the strengths of neural and symbolic techniques offers promising paths to expand generative AI capabilities. Neural models could provide fluid creativity, while symbolic systems enable more robust reasoning and validation. Linked conceptual knowledge graphs can ground generative models with more profound meaning and consistency. The data and compute-intensive nature of generative AI also differentiates it from other techniques. Distributed training across thousands of GPUs/TPUs on petabyte-scale datasets has fueled progress. Lowering costs and expanding access remain challenging. But optimized hardware, compression, and knowledge distillation can help.

In summary, generative AI innovatively leverages deep learning architectures to synthesize novel content. It represents the cutting edge among current machine learning approaches. However, realizing its long-term potential may require complementary advances in symbolic AI for reasoning and grounding. With prudent development, generative models can expand AI capabilities in ways that augment humans creatively rather than replace us fully.

The Promise and Perils of Generative AI

Generative AI holds tremendous potential to transform industries and enhance human capabilities. But it also poses risks that require thoughtful governance to address. Realizing the benefits while mitigating the pitfalls will determine whether these models are a net positive for society.

Generative AI holds tremendous potential to transform industries and enhance human capabilities by automating repetitive business processes to free up employee time for higher-value work. For example, AI agents can handle tier-1 customer support, invoice processing, contract reviews, and other repetitive tasks following clear templates, improving efficiency and allowing refocusing on strategic initiatives. Previously, leveraging advanced AI required extensive resources. Still, generative models like GPT-3 now enable anyone to harness cutting-edge NLP via simple API access, potentially empowering citizen developers, startups, and underserved communities by removing barriers to entry. Generative AI excels at synthesizing novel artifacts and enhancing human creativity in designing products, visual media, music, 3D models, games, and more through approaches like prompt engineering to ideate differently. Models can also generate custom content tailored to each user's interests and context, allowing more relevant personalized experiences across marketing, e-commerce, search, and entertainment by enabling more targeted advertising and recommendations. Processing large datasets to uncover insights manually is infeasible, but generative AI could accelerate discoveries across domains like healthcare, climate science, and physics by automating hypothesis generation and data synthesis.

However, generative AI also poses several risks and challenges that must be addressed, including the bias and toxicity inherited from the internet data used for training. Without controls, models generate racist, misogynistic, and abusive outputs, requiring ongoing monitoring, dataset vetting, and filtering mechanisms. The ability to generate highly fluent, credible text and imagery also raises misuse risks around fraud, propaganda, and scams, necessitating new authentication and provenance tracking methods to combat deepfakes. Automating specific tasks risks disrupting jobs and industries, and communities reliant on displaced work could face challenges without support and retraining, requiring policymakers to adapt worker protections. The black-box nature of models also limits transparency into their reasoning, making explainable AI techniques essential to provide visibility into decision-making. Despite advanced content generation, current models still lack deeper reasoning capabilities focused more on pattern recognition versus higher-order cognition, limiting unsupervised usage. State-of-the-art models also require millions of computing resources to train, with widespread access dependent on efficiency improvements and cost reductions through model distillation and pruning.

To responsibly develop generative AI and mitigate risks, we must prioritize research into model robustness, transparency, and interpretability. Policy frameworks addressing bias, misinformation, labor, and access must evolve to align with emerging capabilities. Organizations need to implement responsible AI principles and governance. If cultivated judiciously, generative models could unlock tremendous progress and prosperity. But we must proactively address the challenges ahead to steer this technology toward equitable and ethical outcomes serving all.

Responsible Development and Deployment

The rise of generative AI represents a technological inflection point requiring thoughtful stewardship to align innovations with ethical ends. Realizing the full potential of this paradigm shift while mitigating risks necessitates concerted efforts across research priorities, policy considerations, and organizational best practices. These measures, rooted in beneficence and justice, can broadly steer generative models to empower society.

A core priority is developing techniques that open the black box of generative models. The machine learning community must invest in explainable AI methods that provide visibility into model reasoning, attributes, and limitations. This transparency helps identify potential harms, refine model performance, and support recourse for impacted individuals. Approaches like attention mechanisms, feature visualization, and concept embedding help determine which inputs a model focuses on for particular outputs. However, more explaining model generalization and failures still needs to be done. Ongoing research into model introspection, formal verification, and standards for explainability is essential to engender trust and perceptibility.

Generative models today demonstrate limited reasoning, brittleness outside the training distribution, and vulnerability to adversarial inputs. Researchers must create techniques that enhance model capabilities while minimizing harm. This necessitates expanding context and grounding in real-world knowledge, detecting anomalous or illogical results, comprehensive and inclusive dataset sampling, and adversarial training to improve robustness. Hybrid approaches combining neural methods with symbolic AI could also overcome limitations. There are open challenges in scaling such techniques while maintaining benefits, but safety and reliability are crucial as capabilities advance.

The rapid pace of progress necessitates reflection on how generative AI should progress – purely chasing benchmarks on narrow tasks risks unintended consequences. The research community should take an expansive view encompassing social impact and ethics. Important frontiers include personalization, creativity, and human collaboration. However, the objectives and measures of success must align with broad societal benefit, not solely efficiency. Setting ambitious, yet responsible, goals will help generative AI flourish equitably.

As generative models grow more capable, alignment with ethical priorities becomes imperative. Initiatives like machine learning safety research and AI value alignment explore techniques to embed human values like fairness, accountability, transparency, empowerment of the disadvantaged, and preserving truth. For example, models that synthesize disinformation or toxic content require interventions to override harmful incentives. Constructing generative systems that respect human dignity and justice is an ongoing challenge requiring multidisciplinary perspectives. But putting ethics at the center is essential for benevolent outcomes.

The data intensity and creative capacity of generative AI pose policy challenges around privacy violations and misinformation. Generative outputs could leverage or expose private data or spread falsehoods that appear authentic. Policymakers should require protections like consent, data minimization, and transparency while encouraging research into authentication methods for generated media. Content moderation becomes more complex with AI-generated text, audio, images, and video. Policies should incentivize platforms to balance free expression with mitigating clear harms.

Automating specific tasks may displace jobs, requiring updated policies to support workers in transition. Proactively assessing generative AI's impact on the labor market and providing job training, continuing education, and income/healthcare support will be critical. However, the net effects could be positive if policies elevate human roles to more meaningful work. Policymakers should adopt worker protections while fostering opportunities for this AI-powered shift to augment quality of life.

Policy can shape generative AI to empower disadvantaged communities rather than exacerbate inequality of access. Governance frameworks should ensure affordable access to resources for small businesses, non-profits, and underserved groups. Policies can also incentivize capabilities that overcome limitations faced by people with disabilities. Making generative AI's benefits inclusive by design must be a policy priority.

Policy can encourage accountability and self-regulation by making ethics central to incentives. Potential options include liability rules that impose penalties for harmful systems, certification programs for responsible AI practices, and public algorithmic auditing. A nuanced policy approach can nurture innovation while steering development towards just outcomes. The goal should be fostering AI that serves humanity inclusively.

Organizations leveraging generative AI should integrate it within established ethical governance programs encompassing risk assessment, principles, red lines, and oversight processes. This enables detecting potential harms early and instituting appropriate controls. Governance also helps formally assess tradeoffs and implement policies like human-in-the-loop reviews before deploying specific applications. Integrating generative AI into robust governance is essential for responsible adoption.

Generative AI offers many promising applications that, thoughtfully applied, can provide social value. Organizations should perform impact assessments to identify and enable use cases that foster creativity, personalization, and access without infringing on human rights and dignity. Generative models for education, healthcare, art, environmental sustainability, and adaptive technologies for people with disabilities are worthy goals if developed considerately.

Through testing and reviews, organizations must rigorously monitor generative AI outputs for harmful errors, biases, and deception threats. To minimize risks, ongoing audits help refine datasets, model architecture, and application parameters. Enabling feedback loops from users impacted by errors is also critical. Transparently tracking and mitigating these issues should be standard practice.

To responsibly adopt generative AI, organizations must provide workforce training and support to adapt to changing roles. This minimizes displacement while enabling workers to leverage AI for productivity gains and more meaningful contributions. Complementing generative AI with uniquely human creativity, empathy, and judgment skills is vital. Managers should ensure workers are partners in thoughtfully deploying these technologies.

Organizations should be transparent with users about when interactions involve generative AI. This builds an understanding of capabilities and limitations, which is essential for decision-making applications. Providing opt-out choices and avenues to appeal erroneous outputs helps safeguard rights. Transparency enables informed consent and bolsters trust in legitimate use cases.

The path ahead requires continuous collaboration across stakeholders to develop generative AI responsibly. But if we learn together, we can create an ethical, inclusive future uplifting humanity.

The Generative AI Era

The rapid advances in generative AI over the past few years foreshadow profound changes ahead as these models proliferate across domains and integrally permeate workflows and applications. Progress will likely remain uneven across industries based on investment, data availability, and receptivity.

In the next five years, anticipate continued expansion of models and applications as capabilities improve, talent and infrastructure grow, and specialization for different verticals emerges. This will move deployments beyond novelty use cases towards more integrated adoption.

Longer-term, generative AI can be immensely transformative but poses risks if not developed thoughtfully. We are unlikely to achieve artificial general intelligence shortly, if ever. But we can expect to see increasingly ubiquitous generation and personalization as the technology matures.

However, risks around misuse, bias, and disruption remain, requiring ongoing vigilance. Realizing an equitable future aligned with human values demands purposeful collaboration across stakeholders. This is needed to prioritize responsible stewardship based on ethical principles, foster inclusive access and development, and shape policies through democratic cooperation.

Such efforts can steer innovations toward justice and empowerment rather than extraction. This critical inflection point requires wisdom and courage to develop generative AI as a profoundly positive force for humanity while addressing the complex challenges ahead through open, collaborative governance, putting principles over profits.

If stewarded judiciously, this new era could catalyze tremendous progress, accessibility, and understanding. But outcomes are not preordained. Our choices today will determine if AI uplifts or undermines human dignity and shared values.

The rapid advances in generative AI over the past few years foreshadow profound societal changes ahead as these models proliferate across domains, permeating workflows and applications. However, progress will likely remain uneven across industries based on varying investment, data availability, and organizational receptivity to leveraging these emerging capabilities.

In the next five years, anticipate continued expansion of generative models and applications as capabilities improve through advances in model architecture, training techniques, and computing infrastructure. Talent development in this space will also expand to meet demand, and specialization of models for different verticals and use cases will emerge. This will move deployments beyond novelty proofs-of-concept towards more integrated adoption, where generative AI streamlines actual business processes and end-user experiences.

Longer-term, generative AI can be immensely transformative if thoughtfully cultivated but also poses risks of misuse, bias, and disruption if stewardship lags capabilities. While we are unlikely to achieve artificial general intelligence on par with human cognition shortly, if ever, we can expect to see increasingly ubiquitous and specialized generation capabilities and personalization. As this technology matures, it may transform industries, enable new creative possibilities, and enhance accessibility through customization.

However, risks remain around generative models spreading misinformation, exhibiting harmful biases that exacerbate social inequities, displacing jobs, and compounding climate impacts due to energy-intensive computing requirements. This necessitates ongoing vigilance and governance.

Realizing an equitable future aligned with human values demands purposeful collaboration among stakeholders across the private sector, government, and civil society. This collaboration should prioritize responsible stewardship of these powerful technologies based on clear ethical principles and partnering with domain experts and impacted communities to monitor for harm. Fostering inclusive access and development is crucial to preventing generative AI from widening digital, economic, and social divides. Policy and incentives should steer innovations toward justice, empowerment of all people, and augmenting human capabilities rather than capital extraction alone.

This critical juncture requires wisdom, courage, and moral leadership to develop generative AI as a profoundly positive force for humanity while proactively addressing complex challenges ahead through open, collaborative governance frameworks that put ethics before profits or technological ambition alone. If stewarded judiciously, this new era could catalyze tremendous progress, accessibility, and understanding. But outcomes are not preordained, and any tool can be misused. Our choices today will determine if AI uplifts or undermines human dignity and shared values. Will we realize an enlightened future guided by our highest ideals? Or will we fail to act with the necessary care? The path ahead remains open, contingent on human stewardship. But this generative moment is rich with opportunity if we have the will to shape it responsibly.

Conclusion

The emergence of generative AI represents a turning point laden with both profound opportunities and risks. These models can synthesize novel, high-quality content, unlocking new possibilities for creativity, personalization, and democratizing access to advanced AI. Applications span nearly every industry, from automating repetitive tasks to accelerating discoveries in healthcare and sciences. But genuine risks require diligent governance, including potential bias, misinformation, job displacement, and lack of transparency.

Realizing the benefits while mitigating the pitfalls necessitates inclusive collaboration to develop ethical frameworks for model governance, safety, bias mitigation, labor impacts, environmental stewardship, and other critical factors. With prudent research and policies aligned with global AI governance standards like the ISO/IEC AI Standards, generative AI could enhance human capabilities, increase the accessibility of information, and provide insights to serve society better. But unchecked proliferation or prioritizing narrow interests over collective welfare will undermine public trust and stunt progress.

The path forward must align generative AI with shared human values of equity, responsibility, and sustainability. This requires dedicated research into robustness, reasoning, transparency, and efficiency to address limitations around brittleness, limited logic, data needs, and computing costs. Policy considerations around privacy, content moderation, job retraining, access guardrails, and incentives for accountability are also essential to steer toward just outcomes. Organizations should integrate models into governance programs focused on responsible use cases, monitoring for harmful errors, retraining workers, and providing transparency to users.

Generative models can profoundly augment human creativity and productivity if developed judiciously through collaborative governance engaging diverse voices. But outcomes are not preordained. The extent to which these technologies uplift or undermine our highest ideals remains contingent on human stewardship. We must demonstrate the wisdom to shape an AI future aligned with justice, human dignity, and care for our shared world. We can unlock immense progress for business and society with open and inclusive collaboration. But we must start making the right choices today, and the most crucial intention is to use global standards to measure the impact of Generative AI systems on humans and the environment.

Dr. Seth Dobrin
Founder and CEO of Qantm AI
New York City, NY, USA

About the Author and the Contributors

Sid Ahmed Benraouane is a thought leader with 20+ years of experience in digital transformation, technology policy, and government innovation. Dr. Sid Ahmed Benraouane leads the AI Practice at Harrington Management Systems (HMS), a Silicon Valley consulting firm that specializes in Excellence, Innovation, and Performance Management. He is an expert member of the US delegation at ISO and has contributed to the drafting of the ISO/IEC 42001 AI Management System as well as other AI standards developed by Working Group 1, "Foundational Standards," of the ISO/IEC JTC 1/Subcommittee 42. Previously, he has led the US working group at ISO within TC 279 which has drafted ISO 56002:2019 Innovation Management System.

Dr. Benraouane has published many books on strategy and technology policy, technology implementation, and innovation. In his book *Using the ISO 56002 Innovation Management System: A Practical Guide for Implementing and Building a Culture of Innovation*, Dr. Benraouane provides a step-by-step planning process that helps government leaders design an ecosystem that engages and empowers innovators and inventive minds.

Dr. Benraouane is a speaker at the World Government Summit (UAE), BAIDEC/MEMTEC conference (Bahrain), and Big Data and Cloud Show (Kingdom of Saudi Arabia). He has taught strategy and business for many years at Carlson School of Management, University of Minnesota, USA. Currently, he is advising the Dubai Government, in the UAE, on foresight, innovation, and digital transformation.

Dr. H. James Harrington is a thought leader and a guru of the Quality and Excellence movement in the world. He has authored more than 60 books on quality, excellence, performance management, and innovation, and has led many awards, including the Asia Pacific Quality Organization (APQO) Award, and the Harrington-Ishikawa Award, a global award that is named after him and the late Ishikawa.

For more than 40 years, Dr. Harrington led complex projects at IBM. He was a principal at Ernst and Young for 10 years and was a member of more than 14 different boards of Directors of successful organizations. He also served in the US Navy.

Dr. Harrington also served as the past President and the Past Chairman of the Board of the American Society for Quality (ASQ) and the International Academy for Quality. He is one of the few people who are licensed by the US government as a professional quality engineer.

Dr. Harrington is a lifetime Honorary President of the Asia Pacific Quality Organization. He was also named president and chairman of the Walter L Hard Foundation. Currently, there are seven performance improvement and student grants named after him. As a US delegate to ISO, Dr. Harrington participated in the drafting of many landmark ISO international standards, including ISO 9000 Quality Management Series, ISO 14001 Environmental Management Series, and ISO 56000 Innovation Management Series.

Dr. Harrington is an entrepreneur with a passion for Excellence and innovation. He founded a consulting company in the Bay area, Harrington Management System, which has been involved in delivering Excellence and innovation programs and building performance capabilities in many industries and countries, including the UAE, where he established a lifelong partnership with the Dubai Government and Dubai Police. His involvement with the UAE government led to the emergence of a new leading Excellence framework, the Government Excellence Model (GEM 2.0), which has become a gold standard for government excellence assessment programs.

Dr. Seth Dobrin, a distinguished leader in artificial intelligence (AI), is the founder and CEO of Qantm AI. With over 20 years of experience, Dr. Dobrin's innovations include his role as IBM's inaugural Global Chief AI Officer, bringing a human-centered approach to AI strategies in complex business landscapes. As

Chief Data Officer at IBM's Software Group, he championed business modernization through data-driven insights. Dr. Dobrin is a respected advocate for responsible and inclusive AI, earning accolades like AIConics Innovator of the Year and recognition among the Top 10 Most Influential Chief AI Officers in 2023. Holding a doctorate in molecular and cellular biology, his multidisciplinary approach guides his work at the intersection of AI, data, and business transformation. He's a champion of diversity in tech, speaking at conferences and industry events and featured in tier-one media outlets like BBC and *Forbes*. Despite his life sciences background, Dr. Dobrin has gained experience across multiple industries.

Introduction and Book Organization

The recent progress achieved by Large Language Models and Generative AI has intensified the calls for the adoption of Responsible AI (RAI). Text and image-based applications, such as ChatGPT, Bing, Bard, Stable Diffusion, and Midjourney, as well as multimodal applications, such as Runaway and Gen. 2, emerged to unleash a new race towards AGI. These breakthrough AI products, and the underlying foundation models that power them, have democratized AI systems, accelerated the social interest in them, and raised an alarm about the need to address their risk and harm to society. Calls from business leaders, technology experts, and civil society advocates, to regulate AI and address its risks and harms have intensified. Even the founders of leading AI companies have called for the need to regulate and audit this new technology. In his congressional panel hearing in 2023, Sam Altman, the founder of OpenAI, has called for a federal agency, modeled around the US federal Food and Drug Administration, or an international agency, similar to the International Atomic Energy Agency, to evaluate Generative AI performance, and license its applications.

The race to adopt AI technology has also accelerated. The global adoption rate of AI technology has reached 35 percent, a four-point increase from a year before, according to the 2022 IBM Global AI Adoption Index. The perception of adoption speed also stands at an inflection point. A survey by KPMG published in 2021 about how fast AI technology is moving, found that 49 percent of business leaders in the tech industry believe that AI technology is moving faster than it should be. In financial services, 37 percent of business leaders think AI is moving too fast, while in industrial manufacturing and the government sector, the perception of speed is at 55 percent and 37 percent, respectively.

However, despite the increasing speed of adoption, companies are significantly lagging in developing robust frameworks for Responsible AI. The implementation of RAI programs remains limited and is still in its infancy stages of development. A global executive survey conducted by MIT Sloan Management Review and BCG, in 2022, found that only 52 percent of organizations have a RAI program in place, with 72 percent of them saying that their RAI programs are still limited in scope and scale. More importantly, 84 percent of executives surveyed believed that they should do more to build a responsible AI strategy.

The gap between perception and reality is a wide one. Many reasons explain this gap in perception. First, technology is simply moving faster than organizations have the time to think, reflect, and design mitigating strategies that protect them from technology liability and empower their users and customers to use the technology responsibly. Second, RAI is a relatively new field and literature about the RAI concept is still developing. Emerging fields tend to struggle before achieving a certain level of consensus that captures the maturity level in concepts and terminologies. And, third, there is confusion around the concept usage itself. The terms *AI ethics*, *trustworthiness*, *responsible AI*, and *AI governance* have been used sometimes interchangeably, thus adding more to the confusion.

Responsible AI refers to the governance framework organizations adopt to put in place an AI model that is responsible. An AI model is said to be responsible if it subscribes to ethical principles, such as fairness, security, safety, privacy, robustness, transparency, and explainability, human oversight, and do no harm principle. A global consensus seems to have been building around these principles, as shown in the frameworks published by leading international institutions, such as OECD, UNESCO, ISO, EU, NIST, and the EU High-Level Experts Group. In the private sector, many leading technology and business companies, such as IBM, Microsoft, Apple, Google, Salesperson, Accenture, PWC, Deloitte, BCG, and EY, have adopted a similar version of these principles and have raised concerns about the unintended consequences of AI and its misuse.

Since 2017 ISO, the International Organization for Standardization, and the International Electrotechnical Commission, through their joint technical committee (JTC1), working groups, specialized committees, and their global network of experts, have published many international standards on AI systems, machine learning, AI use cases, trustworthiness, and other technical procedures to implement AI systems and models. One of these critical and foundational standards is the International Standard

ISO/IEC 42001 Artificial Intelligence Management System, a standard that is meant to provide organizations with a framework to put in place a system of AI that allows the organization to use AI coherently and responsibly.

I have been involved with many ISO/IEC discussion groups. As a member of the US delegation at ISO and a member expert of the group that drafted this standard, I wanted to contribute to the overall conversation about how we can take advantage of the many opportunities AI technology brings to solve crucial problems we face, while at the same time making sure that AI is used wisely.

To help the reader understand the issue and provide guidance to managers interested in implementing an AI management system, I organized this book in the following way.

In the first part of the book, I introduce AI, the history of AI, its typology, its evolution, and its economic impact. While I was writing this book, Generative AI (GenAI) broke into the conversation, so I decided to call on my friend Seth Dobrin, former Global Chief AI Officer at IBM and the founder and CEO of Qantm AI for help. As an expert in AI, Seth is a leading voice in responsible and inclusive AI and the Top 10 Most Influential Chief AI Officers. He generously offered to help me in this book by writing an entire chapter. Generative AI, in which he describes the promise and the challenges of Generative AI.

The second part of the book delves into the heart of the issue of ISO/IEC 42001. Here, I provide my interpretation and my reading of the standard. My goal is to offer a simple interpretation of how you can implement the standard to put together a viable AI management system.

So, in this part I discuss how you analyze the context of the organization and what tools you may use to understand the context with some best practices and examples; I address the role of leadership in the governance of AI with examples of how to develop an AI strategy that conforms to the standard; I analyze the planning process and provide a typology of the concept of risk and what best practices are out there that may help you come up with your risk management framework; I explain, in some details, the type of support organizations need to have to enhance the functioning of an AI management system, with a detailed discussion of AI competency framework; I show how an AI project life cycle ought be run (operations); and finally, I provide guidance on how to run a performance evaluation process, and implement an improvement plan to enhance the functioning of your AI system.

You still have to buy the standard and read it. It is critical to have it as your main source. In the standard, you will also find implementation guidelines (Annex B). These are recommendations coming from the collective wisdom of experts who wrote the standard. Out of respect to these great men and women who volunteered their time to work on this standard, I decided not to discuss their guidelines and leave it to you to make sense of it. These guidelines are one way to implement the standard, your reading would be another way, and my book would be a third way of implementing the standard. Altogether, hopefully, we will be able to develop a sustainable governance framework that helps us take advantage of this technology while using it wisely.

1

ARTIFICIAL INTELLIGENCE AND GENERATIVE AI

Forces behind the Digital Transformation

Chapter 1

Artificial Intelligence: A Transformational Technology

PURPOSE

The purpose of this chapter is to explain the history and the evolution of artificial intelligence. The chapter defines AI and discusses the different types of AI, including the difference between Artificial Narrow Intelligence (ANI), Artificial General Intelligence (AGI), and Artificial Super Intelligence (ASI).

OVERVIEW

The chapter discusses the overall historical evolution of artificial intelligence as a scientific discipline as well as the different historical phases this technology has gone through since the 1950s. The reader will find a brief history of AI evolution, including the main historical era that characterizes the evolution of this technology and a discussion of the main different types of artificial intelligence. The chapter's objective is to underscore how this technology started to emerge slowly as a driving force for change pushing organizational leadership, and society at large to start a conversation about the benefits and the challenges this technology poses to society. A brief presentation of the main thought leaders who shaped the evolution of this technology is also included.

DOI: 10.4324/9781003463979-2

BENEFITS

At the end of this chapter, the reader will be able to:

- Define AI and differentiate between the three main types of AI.
- Understand the impact of this technology as well as its transformational power.
- Learn about the challenges this technology poses to society.

Introduction

Just a few years ago, the topic of artificial intelligence (AI) was a specialization of academic research institutions and tech labs, and speculation of scientific movies such as *The Matrix* and *Her*. Today, AI seems to have moved to mainstream society and has become the center of the discussion of CEOs, business leaders, policymakers, strategists, and entrepreneurs willing to capitalize on this historical shift that will change the way we conduct business during the last 100 years. More than half of CEOs surveyed in 2023 think that AI will have a larger impact than the internet, while 80 percent of CEOs think AI will significantly change the way we conduct business in the next five years, according to reports from McKinsey and PWC.

Andrew Ng, a leading scientist in machine learning, argues that the impact of AI on our society will be similar to the impact of electricity on world civilization. When electricity was discovered and became accessible to everyone, it became the engine that fueled innovation in every single segment of the economy. It became the spark that lit up progress and scientific development in the twentieth century.

As a general-purpose technology, AI is poised to transform the way we do business today. Particularly true for new foundational models, AI will become the critical source of innovation shortly profoundly shaping the global economy. Successful companies will be those who move first on introducing applications based on Large Language Models to take advantage of the disruptive nature of the technology. Economic sectors, such as transportation, logistics, manufacturing, defense, and national security, will be deeply affected by AI, while traditional management functions such as customer service, operations, logistics, HR, finance, and marketing will experience a fundamental shift.

So, what is AI, and how it is different from human intelligence? How will AI affect business and the global economy and what should managers do to take advantage of this technological breakthrough? What will be the impact of this new technology on social norms and how will society react to it? These are some of the questions that we will address in this chapter.

Definition of AI

Before we get into all these questions, let's start with some conceptual frameworks that will help us understand the nature of this nascent technology and how it will impact our lives.

AI is a branch of computer sciences that has been evolving since 1950. The term was first coined by the American cognitive scientist and inventor John McCarthy in a conference paper presented in 1956 at the Dartmouth College Conference. The ambition, at that time, was to create a machine that mimics the activities a human brain performs during the learning process. In McCarthy's mind, by automating different learning tasks a human brain goes through, we may be able to program a machine to learn and behave in similar ways to humans.

Haugeland (1985), an American professor of philosophy and cognitive scientist, provides another definition of AI. For Haugeland, AI refers to "the exciting new effort to make computers think … machines with minds, in the full and literal sense." In this definition, Haugeland believes that AI in the future could acquire capabilities that are similar to the ones humans have. These capabilities are abstract thinking, emotions, feelings, and consciousness.

Kurzweil (1990), an American inventor and futurist, provides yet another definition of AI. He argues that "AI is concerned with building machines that can act and react appropriately, adapting their response to the demands of the situation." In this definition, Kurzweil adopts a behaviorist approach to AI to point out the ability of machines to act intelligently. Figure 1.1 provides an overview of the history of AI evolution since 1956.

Different Types of AI

Because not all of the types of AI we talk about today have the same power and impact, it is important to remind the reader about the distinction that

Figure 1.1 History of AI: the evolution of the AI technology over time.

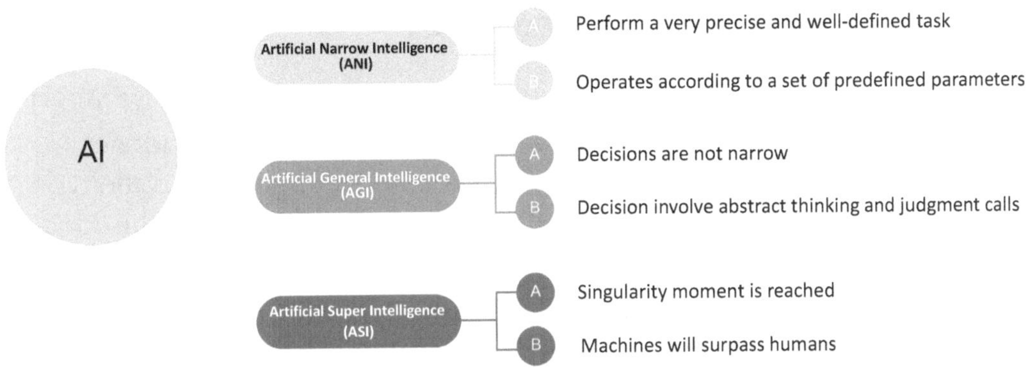

Figure 1.2 Different types of AI.

needs to be made between the three types of AI. In the field of AI, we make the distinction between Artificial Narrow Intelligence (ANI) or narrow AI, Artificial General Intelligence (AGI) or general AI, and Artificial Super Intelligence (ASI) or super AI. Important to mention that at this junction of history, we have been able to discover only narrow AI. We are not able yet to use the general AI or super AI. Figure 1.2 provides a summary of different types of AI.

Artificial Narrow Intelligence (ANI)

This type of AI, also called weak AI, is designed to perform a very precise and well-defined task. It is narrow because it does not go beyond what it has been programmed for. It operates according to a set of rules and a set of predefined parameters. A good example of a narrow AI is the language processing programs we are all familiar with, such as Apple's Siri and Amazon's Alexa. Another example of narrow AI is the AI algorithm behind search engines, such as the RankBrain program Google uses, to map queries, organize searches, and rank pages according to a specific parameter that is provided by the programmer.

These applications, while they appear to be very intelligent in interacting with us, perform in a very narrow fashion. They perform only actions for which they were programmed. Apple Siri, for instance, while it may seem that it knows everything, interacts only from within what has been provided to it by the datasets. In other words, it cannot go beyond the parameters from which it reads. That is why Siri, or any other language processing application, is unable to understand language subtleties such as humor, jokes, sadness, or emotion in general.

Artificial General Intelligence (AGI)

The second type of AI is General AI. This type of AI, also known as full AI or strong AI, has the potential of mimicking human cognition, and the decision-making process that a human brain goes through. This is the type of AI that we see in Sci-Fi movies. While we are still too far from this kind of AI, the exponential nature of technology will one day lead us to strong AI where computers will be sovereign to make decisions outside of human control. This is what we refer to as the Theory of Mind AI, where computers will be able to discern human thinking, human reasoning, and emotions. Equipped with the Human of Mind AI, AGI will be able to contextualize knowledge and understand the world. Computers' decisions will not be narrow. They will involve abstract thinking and judgment calls about situations that are surrounded by ambiguity and uncertainty.

Artificial Super Intelligence (ASI)

The third category of AI is called super AI. In this type of AI, we reach the singularity moment where machine intelligence surpasses human intelligence. In this type of AI, the machine acquires qualities that are so far exclusively part of the human experience: consciousness, a term that characterizes humans and refers to our ability to know, feel, perceive, and be aware of our surroundings. In ASI, computers will not only be aware of their surroundings but can be aware of themselves.

Chapter 2

Generative AI: A "Spark from AGI"

PURPOSE

The purpose of this chapter is to define generative AI and explain its transformational nature. As a breakthrough technology, generative AI has changed the conversation about AI, and how society should address its benefits, as well as its risks.

OVERVIEW

The chapter discusses generative AI, its evolution, and the impact this technology is expected to have not only on business but also on the wider society. The chapter delves into the expected economic impact of generative AI, as well as the most likely economic sectors that will be impacted by this technology. The chapter also presents a review of recent studies that attempted to quantify generative AI economic impact and discuss the challenges, the risks, and the harm associated with this technology. Issues such as privacy, security, and cost associated with generative AI are all discussed.

DOI: 10.4324/9781003463979-3

BENEFITS

At the end of this chapter, the reader will be able to:

- Define generative AI and understand the how and the why of the sudden public interest in it.
- Understand the potential positive impact of generative AI on the economy and how these benefits are shaping the conversation about digital transformation in general and AI in particular.
- Learn how business leaders should prepare for generative AI adoption and what change policy leaders should have in place to prepare for the implementation of this technology.
- Understand the risks and harms of generative AI, and how organizations and society should mitigate them.

Introduction

When Open AI released its ChatGPT application, the conversational chatbot built on the Large Language Model, it ignited a new discussion about the role of AI in our society, and whether the age of AI has truly begun. The new release, presented as "sparks of AGI" by Open AI researchers, was a turning point in the history of AI that has captured the imagination of the world. In less than ten days, ChatGPT became the most popular AI application in the history of the internet, reaching more than one million active users in five days. As the entry barrier suddenly got lowered, entrepreneurs willing to test their ideas, rushed in with new generative AI tools and applications sparking serious conversation among researchers, policymakers, ethicists, and technologists on whether legal authorities should step in to quickly rein in the new technology. Text and image-based applications, such as ChatGPT, Bing, Bard, Stable Diffusion, and Midjourney, as well as multimodal applications, such as Runaway and Gen. 2, emerged to unleash a new race towards AGI. On the venture capital front, investment in generative AI increased by an average compound growth rate of 74 percent between 2017 and 2022 according to a 2023 McKinsey report. New generative AI unicorns, with a valuation of $1 billion-plus, such as Anthropic AI, Cohere, Hugging Faces, and, of course, OpenAI, dominated the VC field as of May 2023, according to the technology research company CB Insights.

What Is Generative AI

Generative AI is a subfield of AI technology that produces various types of content, such as text, image, audio, and synthetic data. Generative AI is a subset of deep learning technology that creates new content that is based on training data. In its current form, Gen AI is still a version of Artificial Narrow Intelligence (ANI or weak AI), the type of AI that performs according to predefined rules and parameters. Gen AI can be general, like the ChatGPT application that answers all types of generic questions, or a specific one trained on a specialized subject matter expertise, such as a chatbot for finance, law, and medicine.

In the acronym GPT (Generative Pre-trained Transformer) the magic of GPT lies in the concept of "Transformer." Transformers are neural network architecture used in Natural Language Processing (NLP). Their extraordinary ability to predict the subsequent words in a standard sentence stems from a mathematical formula that enables them to assign weights to various word options. Since its introduction in 2017, transformers architecture has revolutionized the landscape of NLP. They also called into question our assumptions about how, in philosophy and linguistics, we used to associate language with thoughts. Transformers have mastered a "formal linguistic competence" that teaches them the syntax of a language, and the set of rules and principles that govern the proper arrangement and ordering of words to convey meaning. This formal linguistic competence is what gives transformers their power of knowing and predicting. It is what makes them magic, and it is what inches them closer to a cognitive ability that we thought was the essence of humanity.

As an emerging technology, there are a few things that make generative AI different from previous technological breakthroughs.

First, the ability of this technology to sift through a huge amount of data is beyond what a human can imagine. For a General-Purpose Technology to be able to read data faster, analyze it, categorize it, summarize it, and provide a coherent answer in less than seconds makes this technology a truly transformational one that will have a deep impact on every aspect of our life. More importantly, the ability of generative AI to develop emerging abilities, abilities for which it has not been trained, makes this technology a different one. It truly gets humanity a step closer to AGI moments.

Second, the ease of access and the ability of people of all sorts to interact with this technology is in itself unprecedented. Entry barriers for previous breakthrough technologies were relatively higher. ChatGPT application has truly lowered entry barriers to the point that the fourth-grade-level student can work and interact with generative AI. Even for writing codes and software, generative AI does not require the previous technicities, and labor-intensive work that writing codes and software used to require. As said by Microsoft CEO, if the world has around 100 professional developers now, with generative AI, that number can get to 1 billion software developers, and that is because the entry barriers to this field have been lowered.

TOOLBOX 1

THE PRESIDIO RECOMMENDATIONS ON RESPONSIBLE GENERATIVE AI

In June 2023, a group of more than 100 experts and thought leaders met in the Presidio of San Francisco, USA. The meeting was organized by the World Economic Forum and was meant to be a platform for discussion and debate about how the world should collaborate to enable innovation and Responsible Generative AI. The goal was to come up with actionable items that help organizations, technical experts, and policymakers address the need *for* Responsible Generative AI.

At the end of the meeting, participants issued 30 recommendations. The Presidio recommendations addresses topics that relate to the design and development of generative AI applications, how open innovation and international collaboration should guide innovation, and how society and policymakers should address the adverse impact of the transformation.

Issues like the need to establish precise and shared terminology from the start, how to build public awareness of AI capabilities and their limitation, how to keep the focus on human values and preferences, how to encourage stakeholders' engagement, and how to engage in rigorous benchmarking to uphold accountability in AI research formed the essence of the Presidio recommendations for a Responsible Generative AI. For more information about the summit, visit this webpage:

https://www.weforum.org/whitepapers/the-presidio-recommendations-on-responsible-generative-ai

Generative AI Added Value and Economic Sectors Impacted

In terms of benefits, generative AI is expected to have a deep economic impact. The automation of mundane tasks and the improvement in efficiency will help organizations increase productivity. Knowledge workers will see more impact, positive and negative, from generative AI. As this technology becomes a source of innovation and enhancement to creative and knowledge-based output, organizations will be transformed radically.

According to a KPMG report, 78 percent of respondents to a survey on the impact of generative AI see a high or extremely high impact of generative AI on innovation, technology investment, and customer experience. Generative AI, according to the same report, will have the greatest transformation impact on R&D.

A 2023 report by Accenture on the impact of generative AI reveals that Large Language Models may impact up to 40 percent of working hours across industries, with banking, insurance, and capital markets seeing most of the impact. Job categories and job occupations, such as admin support occupations, sales, business, and financials, will see the most impact according to the same Accenture report. The role of leadership will be critical in providing the direction, guidance, and incentives to transform the organization and reinvent new ways of working where humans responsibly work with generative AI tools.

As to the generative AI market value, and the sectors that will be impacted, an estimate by Franklin Templeton Investment puts the evolution of the AI market at $900 billion by 2026, with a compounded annual growth rate of 26 percent. According to the same report, five sectors will see most of the impact.

- **Healthcare**: The simplicity and ease of use of the generative AI conversational model will democratize access to healthcare services, allowing patients to personalize and customize their care.
- **Technology**: Generative AI will revolutionize Internet searches by allowing users to take control of the way they want to customize their searches, and how to engage with generative AI platforms to get the answers they want. Generative AI will have a deeper impact on technical fields, such as legal, medicine, code writing, and video production, to name a few.
- **Customer Service**: The customer service field will see the most impact of generative AI, particularly for chatbots and conversational AI

tools. The ability of companies to tailor the conversation, provide fast and precise responses, and engage with the customer in a meaningful way is a critical added value of generative AI. AI can sift through large amounts of consumer behavior data, such as demographic data, patterns of clicks, and trends, to identify new audiences or customize content at scale. Generative AI will reinvent how demographic segmentation and audience targeting are being done now.

- **Education**: Finally, generative AI will also impact the educational sector by providing tailored educational materials to students and learners. Educational institutions will need to work hard to find a balanced way to integrate generative AI in a way that does not harm copyright materials and students' learning growth, while, at the same time, providing students with a fun and easy way to enhance their academic experience. Generative AI will play the role of a coach and assistant teaching students about a variety of subjects and topics that support their learning journey. Figure 2.1 provides a summary of the GenAI economic impact.

Generative AI Harm, Risk, and Cost

As pointed out by the Norwegian Consumer Council report, the main issue with generative AI models is that they pose a structural threat to society because they tend to reproduce, in a novel way, existing power structures. By reproducing material that is itself biased, generative AI reproduces harm, perpetuates bias, and masks prejudice that is built into the existing material on which it gets trained. Although it presents itself as "intelligent" and as a

Healthcare	Technology	Customer Service	Education
• More tailored tools for drug discovery • Democratization via the AI conversational model • Better access to healthcare services • Personalization of patient care • New opportunities for biomedical applications.	Gen AI will impact the following technologies: • Product feature optimization • Creation of new AI-based products • New AI-based enhancements of products • Risk modelling and analytics	• New ways to hyper and personalize CX • Interactivity for product search and recommendations • Tailored conversation, for faster and more precise responses • Better CX engagement • New creative ways in audience targeting	• Personalization and enhancement the learning experience • Tailored educational content and assessments • Act as AI tutor to recommend personalized content • Act as a coach and as a copilot to support institutions and students

Figure 2.1 Generative AI economic impact.

system that knows a lot, Generative AI, for the time being, cannot differentiate between right and wrong and lacks any understanding of its own. It is an inherently flawed model, and this is why it is critical to address its risk and harm. As a new and powerful technology, generative AI brings forth numerous societal risks, among them disinformation at scale, hallucination, online children's safety, data breaches, and deepfakes. Let's address some of the risks and harms.

Emerging (and Unknown) Abilities

What sets generative AI models apart are the abilities that emerge over time as a result of the model improving and perfecting itself. These abilities are abilities that have not been planned during the design phase and emerge only when the system starts to work independently. AI experts, such as Geoffrey Hinton, the Godfather of AI, believe that, at least for LLM and Chatbots such as ChatGPT and Bard, the interaction between a large number of different types of features and embeddings gives the model an emerging ability to "understand our world." This ability to understand the world goes beyond the statistical ability of the model to predict word arrangement. And this in itself is a serious risk that makes AI an existential threat. As generative AI develops and as it becomes autonomous in its learning, it inches closer to Artificial General Intelligence.

Harmful Content

The ease of access to generative AI tools makes it possible for bad actors to access these tools to manipulate information and produce all sorts of misinformation that may create harm to society or a specific community. Manipulation, scams, and disinformation at scale could target a country's political institutions such as elections to destabilize the country.

Privacy and Data Protection

Many of the generative AI tools are trained on what is publicly available on the Internet. Some of the material used may be an infringement of copyrights and might impact the privacy of individuals. Pictures, images, and photography, as well as the personal information that come with them, which are scraped from public websites, represent a serious risk for companies using generative AI.

Cybersecurity Threat

The ability of generative AI to learn quickly and develop new capabilities may make it a new threat to cybersecurity attacks at scale. Generative AI can develop new threat models that can be deployed without any technical expertise.

Hallucinations

Large Language Models are known to produce content that does not make sense or is not factually true. The production of this type of content could be a serious threat and could present harm to users.

Finally, and in terms of cost, Generative AI will impact the labor market and the environment. According to the Stanford HAI AI Index 2023 Annual Report, the cost of LLM training is the Large Language Model is very high. LLM models rely on GPU and require a vast amount of data. Hugging Faces applications, for instance, according to its CEO, require the equivalent of 500 GPU of computing power and storage capacity. Version two of Hugging Faces will cost more than $10 million to retrain, while the cost for Open AI to process new prompts that people continuously feed into the application may reach $40 million, according to a CNBC report. Figure 2.2 provides a summary of the risks and harms associated with GenAI.

Emerging (and Unknown) Abilities	Harmful content	Privacy and Copyright Violation	Cybersecurity Threat	Hallucinations
• An ability is considered emerging "if it is not present in small model but is present in large model." • An ability that has not been "planned" for but emerges as a result of the interaction of many elements of the larger model	• Disinformation: creating news and information that looks like true, but it is not. • Bias in the information provided towards or against certain groups • Thought and emotional manipulation	• Some of the LLM are trained on materials that was scrapped from public sites. • This might be a violation of IP and copyrights. • The use of personal information and private data is also a violation of current IP laws and regulations.	• The ability of LLM to facilitate cyberattack • Gen AI as an enabler of cyberattacks. • Gen AI complex algorithm makes it a challenge to figure out the ABCs of attacks • Use of company information as a prompt can be harmful.	• Errors and nonsensical answers you get from your prompt • The answer you get is derived from the training data. It is an extrapolation from information the LLM was fed during the training. • It is an issue because it affects the credibility of the content Gen AI provides.

Figure 2.2 Generative AI harm, risk, and cost.

TOOLBOX 2

FOUNDATIONAL MODELS, FAIR USE DOCTRINE, AND COPYRIGHT ISSUES

One of the issues that Foundation Models has brought to the conversation is the question of intellectual property and copyright law infringement. Data deployers and model creators use publicly scraped data from the internet to train their models. The issue at stake is whether the data scraped from public websites, such as books, articles, writings, and images, should be off-limits. Scraping technology deployed at scale has become a major legal, ethical, and technological risk for companies using these technologies to build models.

In the United States, the fair use doctrine has been utilized to allow the use of copyrighted materials when the use is considered "fair use." According to a study by Henderson et al. (2023), courts employ four criteria to determine the fairness of using copyrighted materials:

1. Transformativeness: Merely using or duplicating copyrighted materials may be seen as a violation of copyright law. However, if the same materials are used with the addition of insights, comments, criticism, or creative changes that transform them, it may not be considered a violation and could be deemed a fair use case.
2. Effect on Markets: A transformative work used for non-commercial purposes may be protected by the fair use doctrine. However, if the transformative work is used for commercial purposes, it may not be protected and could be considered a violation of copyright laws.
3. Nature of Copyrighted Work: Not all materials qualify for the fair use doctrine. Courts consider the nature of copyrighted work to determine whether it qualifies for fair use. Factors such as whether the work is factual or creative, whether it has been published or not, and the level of creativity involved are taken into account when deciding on the applicability of the fair use doctrine.
4. Amount and Substantiality: The final criterion used by courts to assess fair use is how much of the original work has been used and how substantial that portion is. If only a minimal quantity of the original work has been used without permission, it may be allowed under the fair use doctrine. Courts employ both quantitative and qualitative analysis to determine the eligibility of the work for the fair use doctrine.

Foundational models face these issues today, and it remains uncertain whether the data used to train models qualifies for fair use doctrine. In June 2023, a class action lawsuit against OpenAI, filed in California on behalf of six clients, alleges that OpenAI's data collection practices violate copyright laws. A similar case was brought to the court in February 2023 by Getty Images against Stability AI for allegedly violating the copyrights of more than 12 million images.

Source: Henderson, P., Li, X., Jurafsky, D., Hashimoto, T., Lemley, M. A., & Liang, P. (2023). Foundation Models and Fair Use. Stanford University.

Chapter 3

Economic Impact of Artificial Intelligence

PURPOSE

The purpose of this chapter is to discuss the economic impact of AI. The chapter outlines the economic sectors that will benefit from AI as well as the management functions that will most likely see a change as a result of AI adoption.

OVERVIEW

The main discussion thread of this chapter is the economic impact of AI and the transformation this technology will have on the economy. Sectors, such as robotics, manufacturing, finance, and transportation, are presented as an example of where the efficiency introduced by AI applications will have the most impact. The chapter also discusses the current state of AI adoption, as well as the impact of this technology on the labor market and society in general. A brief discussion of how society will react to the rapid adoption of this technology is also presented.

BENEFITS

At the end of this chapter, the reader will be able to:

- Understand the economic impact of an AI.

DOI: 10.4324/9781003463979-4

- Identify the sectors that will benefit the most from AI and the rapid automation of the economy.
- Understand how to use AI in certain management functions, such as customer service and sales.
- Learn about the frameworks and tools used by leading companies during the implementation of their AI technology.
- Identify the opportunities, but also the challenges of AI adoption and implementation.

Introduction

Today the interest in AI is large because of the advances in computing, big data, and the availability of large datasets with affordable access to cloud computing. Peter Diamandis, an entrepreneur and a co-founder of Singularity University, argues that the sudden emergence and interest in AI today is the result of a deep disruption that started to shape up two decades ago. The explosion of many different exponential technologies at the same time, as he argues, created a disruption that has led to the AI revolution. He called it the "unexpected convergence of consequences."

Furthermore, the scientific breakthrough of foundational models made AI an accessible technology to a wider segment of society. The computing power of today's processors, the advances in sensors and networks, the relatively easy access to cloud computing, advances in material science, and breakthroughs in Large Language Model have accelerated the AI revolution and made it more accessible to a large spectrum of economic agents. As an example of the new computing power used in generative AI, Anthropic can process up to 100,000 tokens of text (parts of words or words) for up to 75,000 words per minute which requires tremendous computing power.

Most of the studies that looked at the impact of AI on the global economy agree on the fact that AI is a game changer. The automation of tasks that used to be performed by humans, such as the maintenance of transportation fleet, the management of loans and the detection of fraud, and customer experience, will be the new opportunity that companies have to seize on to capitalize on the power of AI. In a study conducted by Accenture in 2018, it was predicted that AI would increase productivity in the United States by 40 percent, resulting in a net improvement in GDP growth from 2.6 percent to 4 percent. Globally, the AI market was valued at $136 billion in 2022. It is estimated to expand at an annual growth rate of 37.3 percent from 2023 to 3030 according to the World Economic Forum.

Furthermore, AI will not only impact productivity but also will be an important source for the development of new products and services that will have a direct impact on the performance of national economies. In the United States, AI is poised to deliver more than $15.7 trillion in economic growth by 2030 according to a PWC study. Other countries that make up the top 50 percent of the global economy will also see similar improvements. Sweden will see an increase of 37 percent, Finland 35 percent, and Japan 34 percent in their GDP, according to the same study.

Generative AI is also expected to have a significant impact on productivity. A 2023 McKenzie report shows that generative AI could add the equivalent of $2.6 trillion to $4.4 trillion. For the UK economy, for instance, generative AI could add between 15 percent and 40 percent of economic values, which will increase the impact of AI by 15–40 percent.

Banking, high tech, and life sciences will be the most impacted sectors of the economy, while functional management such as customer operations, marketing and sales, software engineering, and R&D will be the most impacted areas of an organization, according to the 2023 Mackenzie report.

Economic Sectors That Will Be Impacted by AI

As we transition from the current economy to an AI-powered economy, most economists agree that the first sectors that will reap the benefits of AI are manufacturing, healthcare, banking, transportation, cybersecurity, and national defense.

The Manufacturing Sector

In manufacturing, for instance, optimization will be the most important outcome of AI. Cohen (2017), from the Economic Strategy Institute, argues that AI will impact three areas: assets optimization, fleet optimization, and network optimization. With AI, manufacturing facilities can become better at managing their performance and optimizing their outputs.

The Finance Sector

In the finance sector, AI is poised to have a major disruption. According to *Forbes* (2018), the banking industry is expected to save more than $1 trillion by 2030. Areas such as risk assessment, fraud detection and management, financial advising service, customer experience, and trading will take more advantage of AI technical advances. Automated algorithmic trading can process a huge amount of data and learn from the pattern of trading to make suggestions for stock trading while watching for triggers to identify anomalies. Similar applications are also used in loan processing and loan management. AI and automation of loans can result in many advantages. It can reduce credit loss, lower due diligence costs, and increase orientation revenue.

The Transportation Industry

In the transportation industry, the AI revolution is already happening. From driverless trains to self-driving cars, AI breakthroughs are expected to change the way consumers perceive transportation. Currently, AI applications are being tested on traffic control systems to eliminate jams, reduce waiting time at traffic lights, and synchronize the flow of vehicles in intersections.

National Security and Law Enforcement Sector

Another economic sector that is considered critical to the US national strategy is the national security and law enforcement sector.

Facial recognition applications have improved tremendously, and now they can be used to monitor traffic, detect suspicious behavior in major events, and predict when a crisis will hit in the future. Predictive policing, an AI application that runs in some US police departments, can now analyze crime areas and map out crime hotspots to predict where crime is more likely to happen. Predictive policing is based on algorithms that learn from historical data to make forecasts about the spike in crime in a specific region.

The Healthcare Sector

In healthcare, AI will be a true game changer. The ability of AI to read images and identify the discrepancy in the image with a higher level of accuracy than a human being will have a deep impact on the economics of this industry. Patients will benefit greatly by getting an accurate reading of their medical images. In a study conducted by Siemens, deep networks were able to detect tuberculosis with an accuracy rate of 97 percent, and a specificity of 100 percent, thus surpassing the human rate set at an average of 60 percent. The reading speed and the accuracy level will be of great help to hospitals producing large amounts of data. Estimates put the data produced by hospitals, in the forms of X-rays, CT scans, MRI, and medical imaging, at around 50 petabytes of data per year. Only machine learning algorithms can process this amount of data and provide a meaningful reading of this huge amount of data to doctors to make an accurate diagnosis for patients.

The Cybersecurity Sector

Finally, cybersecurity is another sector of interest to companies where AI can be critical. AI is more effective with a large amount of data that runs in large networks. AI applications can now spot better and more efficiently malicious activities and can identify intruders that cannot be caught by conventional signature-based detection software. The ability of AI to study and detect patterns that are suspicious and learn from their behavior is what makes AI a powerful tool in protecting national assets and critical infrastructure. Figure 3.1 provides a summary of economic sectors that will be impacted by AI.

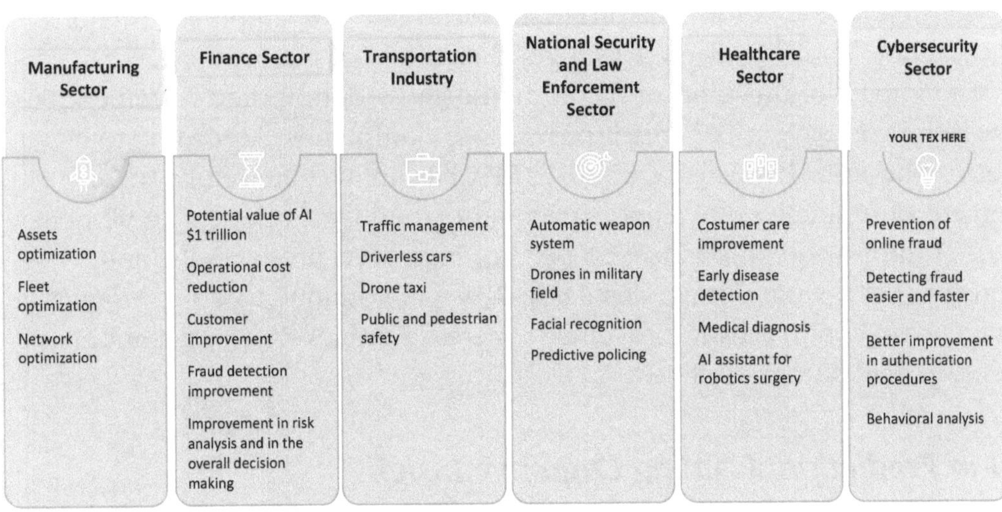

Figure 3.1 Economic Sectors Impacted by AI.

Strategy Implications: The Current State of AI Adoption

A quick survey of AI users shows that worldwide AI adoption is getting some attraction. In a survey conducted by McKenzie in 2021, nearly half of the respondents said that they have integrated at least one aspect of AI and machine learning into their standard business operations. Most of the adoption was in service operation, optimization of new AI-based enhancement products and contact centers according to the same McKenzie report.

A study conducted by Davenport and Ronanki in 2018 reveals that out of 152 companies using AI, 71 use the technology for process automation, 57 use it for cognitive insights, and 24 use it for cognitive engagement. While process automation seems to be the obvious and easy application of AI, cognitive insights and cognitive engagement are the future of AI and hold more promising opportunities. Companies willing to understand market behaviors and connect better with customers to enhance and improve the overall customer experience will find a great strategy in AI.

AI in Automation: RPA

RPA represents the most basic form of AI. In general, any rule-based process has a digital input and routine can be automated. Companies that use RPA use it for back-end office operations such as reading and transferring emails and file, transferring data, toggling systems, applying business rules, orchestrating business processes, executing ERP instructions, and recording keystrokes. This type of AI is programmed to execute simple instructions and cannot "think" by itself. It is a good example of a narrow AI.

The impact of this kind of AI on the business bottom line is tremendous given the cost-saving benefits it provides to companies. According to a study conducted by Deloitte in 2018, RPA has an average return of 20 percent of full-time equivalent capacity provided by a robot, with 92 percent of improved compliance, 86 percent improved productivity, and 59 percent cost saving. Companies using RPA will continue to do so primarily to focus on continuous improvement, increase the level of automation, and develop analytics capabilities.

AI in Prediction: Gaining Cognitive Insight

The second type of AI implementation is cognitive insight. According to the study conducted by Davenport and Ronanki, many companies use AI as a tool to understand better their customers and their markets. In this type of application, companies use AI as a predictive tool to understand users' behavior, detect patterns of clicks, and be able to predict how customers react to different campaigns. According to Davenport and Ronanki, companies using AI use it to achieve the following goals:

- Predict future customers' choices in terms of what product or service they will acquire.
- Understand how to personalize ads to target specific customers.
- Use data on warranty or fraud to decide about trends and what type of customers are likely to benefit from a specific service.

AI and Cognitive Engagement: Enhancing Customer Relationship Management

The third category of AI adopters is companies using AI to engage with their customers to enhance personalization and impact the customer experience. Bots can identify clues during the conversation with a customer to make suggestions or custom products to the specific needs of the customer. Spotify, a music start-up, uses AI to customize music offering to customers based on their preferences, tastes, and habits. Spotify app is one of the most powerful customer engagements as it gets better by the day at predicting customer moods and tastes with high accuracy. Another example of companies using cognitive engagement is China Merchant Bank, one of the largest credit card companies in China. The bank uses the Chinese application WeChat Messenger to handle more than 1.5 million customer conversations a day. These conversations are about banking products, balance inquiries, and payments.

Another example of AI a successful AI use case is the partnership KFC China has established with the Chinese search engine Baidu. In this use case, KFC uses facial recognition applications to customize menus based on the facial expression, gender, and mood of the customer walking into the store.

AI and Robotics

Robotics is an interdisciplinary branch of engineering and science that includes mechanical engineering, electronic engineering, information engineering, computer science, and others. Robotics deals with the design, construction, operation, and use of robots, as well as computer systems for their control, sensory feedback, and information processing. These technologies are used to develop machines that can substitute for humans and replicate human actions.

A robot is a mechanical or virtual artificial agent, usually an electromechanical machine that is guided by a computer program or electronic circuitry. Robots may be constructed to take on human form, but most robots are machines designed to perform a task with no regard to how they look.

Of all the applications of AI, robotics seems to be the one that presents more promise and excitement for the business world. Because of the expected value that programmed robots will deliver, especially in manufacturing, the excitement about robotics has grown exponentially. While we have lived with robots for a long time, the ability to program robots to be autonomous presents a historical moment that gets everyone excited from business leaders to average people.

From an engineering point of view, a robot has two parts: (1) the mechanic infrastructure, which refers to the electronics, the sensors, and the navigation system that the robot needs to move around and gain agility. (2) The cognitive dimension, which is the dimension that helps robots think, reason, and make a decision. This dimension is important because it helps us connect robotics to AI. Many believe that AI and robotics are two different fields and should not be mixed. While this was true for the old generation of robots, which operated under the "rule-based" theory (If/Then), it is not true for the current generation of robots using neural networks. Neural network-based robots do not follow strict rules but decide on their own based on the data, the examples, and the training models they are fed.

Industrial Robotics and AI

The efficiency gains in the manufacturing industry achieved in the last decades are due essentially to the adoption of robotics. In assembly lines, robots execute preprogrammed tasks, such as material handling, welding, and painting, and help companies reduce human errors, produce better quality products, and reduce time delivery. Three factors are driving the interest in robotics: the falling prices of robots, the ability of companies to seamlessly integrate robots into manufacturing, and the ability of robots to perform complex tasks while still collaborating with people. According to a 2022 survey conducted by the International Federation of Robotics, robotics sales grew by 31 over 2020. The main markets for this growth were the United States, China, South Korea, Japan, and Germany. China accounted for more than 51 percent of total robots installed in 2021, reflecting the nature of the economic transformation China is undergoing. In the United States, the field of robotics has gained momentum as companies deploy more robotics capabilities at a higher speed. According to the International Federation of Robotics, this trend, if it continues, will increase robotic density in the United States which is set at 164 industrial robots per 10,000 employees, well above the global average of 74 industrial robots by 10,000 employees. This trend will also reverse the outsourcing movement as it becomes cheaper to use robots on American soil.

Medical Robots

The introduction of robotics into medicine and healthcare is one of the most important AI breakthroughs. While robots have been used in the medical field for the last 25 years, mainly as a tool for guided precision and repeatability, new developments in this technology are revolutionizing the healthcare industry. Robots now in hospitals can perform many functions to support doctors and nurses. Robots can take blood samples, transport analyses, and store them. They can easily detect the right vein when drawing blood, thus reducing stress and anxiety for vulnerable patients. They can also carry carts and beds in large facilities and can even help physical therapists administer their physical therapy.

Military Robots and AI

Military robotics aim at helping soldiers on the battleground to overcome natural obstacles, difficult tasks, and dangerous missions such as walking around minefield or deactivating explosives. Military robots are programmed to operate in unstructured areas and aim to replace humans in dangerous situations. Most of the military applications of robotics tend to be programmed robots controlled by an operator and equipped with image recognition software to help robots identify and adapt to what they see. A rover, for instance, can map out the terrain and check the type of obstacles against its database to choose the right path in case of an obstacle. In the United States, the most known military robot is the Big Dog robot developed by Boston Dynamic and Crusher developed by DARPA and designed by Carnegie Mellon's National Robotics Engineering Center (NREC).

Another military application of robotics and AI is in the field of autonomous weapon and weapon targeting, an application that combines AI and

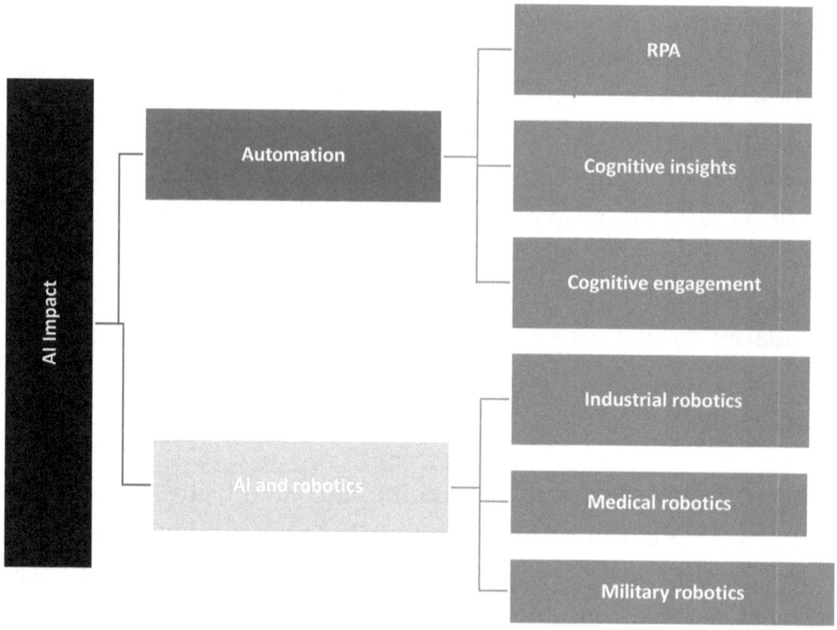

Figure 3.2 Strategy implications.

robotics to train robots to use weapons autonomously. According to Daniel Faggella, who leads the Emerj Artificial Intelligence Research group, autonomous weapons today can identify targets in the specific area they have been assigned to and autonomously track them. This AI application is one of the most important breakthroughs in the defense sector. And while today there is no autonomous weapon that can fire on its own and without clear and explicit instructions from a human agent approving the fire, this type of autonomous weapon will one day be developed. This is why it is important to develop a strong code of ethics that guides and regulates this type of weapon. Figure 3.2 provides a summary of AI's added value.

TOOLBOX 3

WHAT IS THE FUTURE OF TECHNOLOGY?

As AI Spreads, Experts Predict the Best and Worst Changes in Digital Life by 2035

A canvassing report about the best and worst changes in digital life by 2035, published by the Pew Research Center, points out many benefits and challenges that the AI transformation will have in the future. The Pew Research Center, which gathered many AI experts, identifies numerous benefits but also several challenges. Some of the benefits experts have identified include improvements in healthcare and education, better access to knowledge and information for humanity, enhanced connections between people, as well as progress in climate protection and pollution prevention. However, experts have also identified various challenges, including the following ones:

- Harms to the development of digital tools and systems: Innovation and technology development will continue to be driven by profits and economic power, which may result in them becoming controlling tools rather than liberating ones.
- Harm that affects human rights: As digital tools become more powerful, privacy and human rights will be increasingly impacted. Sophisticated surveillance tools, facial recognition, and biometric markers will deeply affect human rights and civil society. Additionally, job displacement will create a greater divide between the haves and have-nots.
- Harm to human knowledge: Basic knowledge and the most reliable information will be at risk due to deepfake technology, misinformation, and the vast production of data. This will affect how people perceive facts and reality, with conspiracy theories potentially influencing people's learning and connections.
- Harm to human health and well-being: The impact of information overload will lead to increased anxiety and depression, isolating people and reinforcing their echo chambers. This will negatively affect their well-being and connection with other community members.
- Harm to communities, governments, and institutions: According to the Pew report, experts believe that political and regulatory institutions will struggle to keep up with the rapid pace of technological advancements. This situation will enable rogue actors to take advantage of the lack of regulations, norms, and standards, exploiting technology for non-legitimate purposes.

Finally, the Pew Research report highlights that 42 percent of experts express equal excitement and concern about the future of technology, while 37 percent state that they are more concerned than excited about the future of AI and technology.

Source: Anderson, J., & Rainie, L. (2023). As AI Spreads, Experts Predict the Best and Worst Changes in Digital Life by 2035. Pew Research Center.

Impact of Automation on Society: How Will Society React to AI and Automation?

When a new technology emerges, there are a lot of unknowns. In the case of AI, the discussion so far is centered on fear, excitement, and hype. While there are a lot of studies that show the positive impact of the new technology on society, there are also calls for the need to be careful in the way we adopt and use AI. So far, the research conducted on the subject of AI falls into two categories. Those who believe that automation will have a positive impact on the economy, and the wider society, and will create more opportunities for people, and therefore we need to embrace it. Those who take a cautious perspective on AI and think that AI will increase job loss and will displace entire segments of the economy, in addition to the risks that it will produce like in autonomous driving, autonomous weapons, and AI-powered cyberattacks.

So how does society react to AI? And what will be the impact of AI on society? How do people react to automation, and would they accept it or reject it?

In addressing these questions, the World Economic Forum (WEF 2019) presents a framework for policymakers with three possible scenarios.

Scenario One: Society Will Accept AI

In this scenario, and according to the WEF, people will accept automation because automation may not develop to its fullest. Automation may simply stumble against some technical obstacles such as the quality of data and the lack of computing power. In this scenario, we face limited progress in automation that will put a break in the momentum and the power of AI. In the labor market, for instance, it will not be possible to automate every single aspect of a job family because jobs tend to have a high level of task diversity, according to WEF. It will be difficult for companies to automate an entire spectrum of tasks. For instance, it will not be possible to automate the job of a financial analyst because, in addition to the repetitive task a financial analyst conducts, he or she also performs other tasks that are based on judgment, creativity, and communication. The complexity that is inherently built into the nature of the job makes it simply difficult to fully automate.

Scenario Two: Society Will Reject AI

In this scenario, society may reject automation because of the irrational behavior that tends to characterize human behavior. In this scenario, the WEF argues that humanity may not be interested in full automation, not because automation is not good, but because, as a human species, we are not always rational in our choices. Even though automation may be the right path for us, as human beings, and because we do not always follow rational paths, we may simply choose not to fully automate our lives. The irrational aspect of our behavior may simply make it difficult for automation to take place thus forcing us to make a selective choice between what we want to automate and what we don't want to automate. As well argued by WEF, much of the support of this scenario comes from the newly produced research in the field of behavioral economics.

Scenario Three: Society Will Accept Automation

The third scenario described by the WEF on the spectrum of acceptance/rejection of full automation is called "planned automation." In this scenario, humanity may reach full automation by developing a rational process for data collection and machine behavior with a clear definition of what is exactly needed. Policymakers can define the ethical and moral framework that will govern this process of full automation. In this scenario, the decision of whether to automate or not, and to what extent we should automate, has to take place through discussion, iteration, and consultation about what society needs and expects from full automation. As well explained by WEF, the government may provide more help to those who may be adversely impacted by the full implementation of automation by providing incentives, such as training programs for displaced workers, ethical guidelines on when AI can be used, and tax breaks to help local governments and disadvantaged communities take advantage of full automation.

The Jobs AI Will Create

As with any digital transformation, AI's impact on the job market will also be positive. While researchers agree that AI and robotics will eliminate many jobs, they also agree that AI will create other types of jobs. Those who are poised to benefit from this disruption will be companies that are moving simultaneously on three fronts: changing and reinventing the business process, rebuilding the human capital of the organization, and utilizing data by intelligently capturing its power.

According to a study conducted by Accenture 2018, AI will create new jobs that we have not seen in the past. These jobs are the jobs of trainers, explainers, and sustainers.

Trainer

According to Accenture's report, the first category of jobs needed is the people who train AI on how to respond adequately to input from customers. There are two types of training needed. Those who will help natural language processors and translators understand the meaning of input on the one hand, and those who will train algorithms on how to imitate human behavior on the other hand. Examples of this category of jobs include training AI on language subtleties, recognizing sarcasm, and identifying the true meaning of a statement.

Explainer

When companies deploy new algorithms, the result sometimes may not be what the company has predicted. In many cases, AI behaves in a way that is not programmed. What we refer to as the "block-box problem" could be solved by training people on how to explain the decision made by AI and make sense of it. Take the example of a bank's algorithm that was built to depict the creditworthiness of customers so the lender can make a judgment about the loan approval process. If the algorithm starts to reject a specific segment of customers, then the lender is expected to explain the why of this situation. Explainers are an important category of the AI value chain because they provide AI with the credibility it needs when it is deployed for the first time. Explainers help business leaders make sense of AI behavior and provide an answer to the public. In Europe, the need for "explanation" has become a legal requirement. The General Data Protection Regulation (GDPR) recently implemented requires companies to explain their decision.

Sustainer

The third category of new jobs that AI is expected to create, according to Accenture, is the type of jobs that will help companies keep AI in check.

In their research of more than 1,000 companies, Accenture found that less than one-third of companies trust their AI systems with high degree. This is a challenge if companies still don't have full confidence of this new technology. Sustainers will need to understand the purpose of the AI systems implemented and the mission of the organization as well as the type of stakeholders. Their role is to watch for behavior that could go wrong, much more like a compliance officer or ethical committee tracking algorithm's behavior.

Countries' AI National Strategy

Because of its transformative nature, AI has become a powerful economic tool that will provide countries with a global leadership position. In the last five years, many countries have launched different documents that outline their AI strategy. Some of these countries are Canada, France, China, the UAE, and the United States. The underline thread for these documents is a political will to organize the countries' resources and capabilities into a coherent framework that will spur innovation around AI and other exponential technologies. While these strategies are different from one another, their common objective is to create an AI ecosystem that targets funding, R&D, education, and private–public partnership. Some of these countries, like China, have also set large budgets to encourage investments in infrastructure and digital transformation, a *sine qua non condition* to AI growth.

In the United States, most observers agree that the US government has not done what it needs to do to organize resources and rally the nation behind this new technology. What the United States has done, in terms of policy, was the publication of a collection of documents that were produced in 2016 and 2023 to lay out the national AI agenda. The first set of documents was released by the White House Office of Science and Technology Policy in 2016, and the latest one was the AI Bill of Rights issued in 2023. All of these documents emphasize the need to develop a long-term investment plan in AI research and the prioritization of investment choices by the federal government. These documents also call for the need to set governance standards and framework, as well as the investment in building an AI workforce.

Chapter 4

Digital Transformation: How to Prepare Your Organization for Change

PURPOSE

The purpose of this chapter is to discuss digital transformation. The chapter outlines the importance of setting up a framework to manage digital transformation and presents some best practices to lead the transformation of the organization.

OVERVIEW

The chapter presents some useful digital transformation frameworks that have been implemented by leading companies. The goal is to help the reader understand what digital transformation is, what it entails, and how complex it can be. Using simple frameworks and tools this chapter argues for a comprehensive approach to managing change and transformation while emphasizing the role of leadership in leading the transformation. Using data from leading research organizations, this chapter outlines a step-by-step process that helps the reader understand how to successfully implement a digital transformation strategy.

DOI: 10.4324/9781003463979-5

BENEFITS

At the end of this chapter, the reader will be able to:

- Understand the concept of digital transformation and how it should be applied.
- Develop a comprehensive framework to transform the organization that includes technology, people, and culture.
- Engage the reader in a conversation about digital transformation, and how ISO standards should be implemented.
- Identify some best practices from leading organizations on how to implement a digital transformation framework.

Introduction

The concept of digital transformation refers to a wide organizational effort senior leadership takes to steer the organization through technology disruptions. It is a process in which the organization uses digital technologies to transform the business model to adjust to market changes. The goal is to help the organization define the business value, articulate the vision, and build an agile capability that helps people reimagine the business, navigate the challenges, and learn how to transcend the traditional role imposed by organizational boundaries.

A study conducted by IDC in 2023 on the importance of digital transformation found that 82 percent of senior leaders agree on the fact that investment in digital transformation is a must. Without investment and a clear prioritization of innovation and digital transformation, companies will be left behind the report argues. An increase in revenue (38 percent), focus on innovation percent (35 percent), and improving operational efficiency (30 percent) are among the top priorities of senior leadership according to the IDC study.

However, a common pitfall for leaders is to think of digital transformation as a transformation that involves the implementation of additional technologies. Companies that fail in digital transformation fail because their approach to digital transformation focuses on acquiring the technology using a plug-and-play approach while ignoring the fundamentals of a successful large-scale enterprise transformation. Ignoring the impact of digital transformation on the overall organization, its performance, and its impact on different stakeholders, as well as how the integration of the acquired technology impacts existing technologies, is a mistake that can be costly to organizations.

A study conducted by BCG in 2020 found that less than 30 percent of digital transformation efforts succeed. A similar study conducted by McKinsey in 2018 points out similar challenges, especially when the transformation involves digital technologies that transcend the traditional role of the IT department. Technologies such as generative AI, deep learning, analytics, cloud computing, and mobile technology have changed the nature of digital transformation.

So, what are the building blocks of digital transformation and how do companies build them? Why do companies fail to build digital transformation capabilities, and how they should acquire them?

Digital Transformation Framework

Since the launch of ChatGPT, the world has changed dramatically. AI, and particularly generative AI, dominates the conversation about digital transformation and is influencing the public discourse about policy and strategy-making. AI has become a powerful tool to transform organizations and a catalyst for change, reshaping people's views, and perceptions about how to conduct a change strategy. Change management literature seems to agree on the need for organizations going through a transformational initiative to address five key issues: Leadership commitment, talent management, innovation and technology, customer centricity, and agility. In the following section, we will delve into these key elements to explain how organizations cultivate and execute them to acquire capabilities that enable them to confront the future and navigate through disruptions. These steps and elements are also summarized in Figure 4.1.

Leadership Commitment: Building Digital Leadership

The presence of a leader who knows about technology and has experience in engaging people and managing transition is a critical building block of the digital transformation strategy. Hard and soft skills with a good understanding of the technology are critical skills. As digital technologies seep into all facets of the organization, having a technology leader who understands how to extract value from technology, while at

TOOLBOX 4

DIGITAL TRANSFORMATION FRAMEWORK

According to a 2020 McKenzie study, leaders who succeeded in digitally transforming their organization to prepare them for AI technologies have excelled in five areas: Strategy, data, technology, and talent and culture.

Strategy	*Data*	*Technology*	*Talent and Culture*
A successful digital transformation starts with a clear strategy on where the company needs to be in terms of technology for the near mid-term and long term. The North Star discussion that senior leadership needs to engage the team in is a must. This in turn helps the team align with the broader goals of the organization.	A data strategy that defines how data is collected, managed, and integrated into the business is a prerequisite of a successful digital strategy. A data strategy helps your manager define the process by which you store and process data. The backbone of digital strategy is a data strategy that is comprehensive, exploitable, and provides directions on where the company needs to be in terms of market and customer choices.	Successful companies in AI transformation invest in building or acquiring technologies, tools, and models that help them leverage the existing capabilities. The goal is to create a powerful framework that guides decision-makers and coaches employees in their day-to-day activities.	According to the McKinsey report, successful companies that have implemented AI transformation strategies have invested in developing internal capabilities and integrated AI technologies into their processes.

the same time understanding how to drive its implementation, is a *sine qua non condition* of success. In the case of AI, leaders who have a good understanding of AI evolution, its business impact, its technical requirements, and how fast it moves, have a better chance to succeed in securing a good transition.

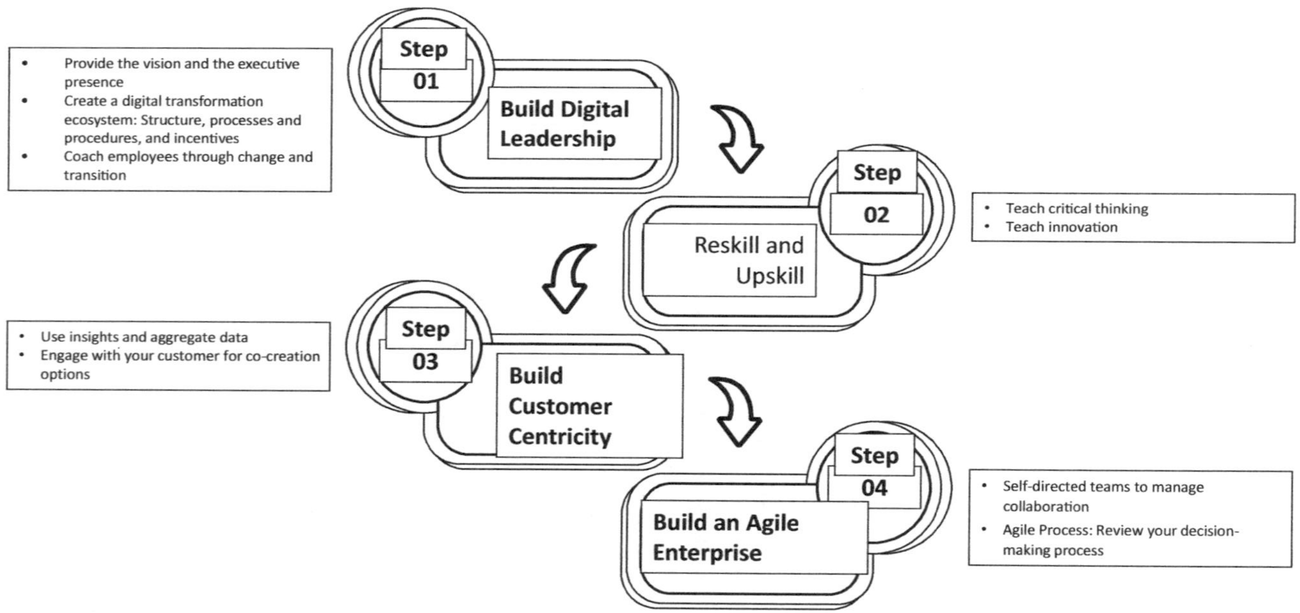

Figure 4.1 Digital transformation framework.

Furthermore, understanding the emerging non-technical requirements is critical to a comprehensive implementation of digital transformation. Understanding the social and ethical implications of AI for instance, as well as the legal challenges different AI applications pose for the company, in the labor and the consumer markets, provides an edge for success. Leaders who understand the nuances of these challenges are in a better position to tackle design issues, as well as implementation issues. These leaders can conceptualize long-term strategies that are nimble and agile. This long-term vision shields the organization from the consequences of changes in the regulatory, social, and political environment. When leaders understand well the issues of privacy and bias in algorithms, for instance, they are more inclined to take a comprehensive and long-term view of the technology. Rather than rushing into quick fixes using a plug-and-play mode, leaders who have a good grasp of the technology build a wider coalition with a variety of stakeholders and work with the technical and business teams to understand the implications of the technology on the customer and the employee.

In a study conducted by Adeva, an HR company specializing in IT talent management, it was found that for digital leaders to succeed in transforming their organizations, they need to develop a variety of skills that go beyond their initial training and specialization. They need to be innovative, be willing to challenge existing norms and take risks, as well as be able to take bold action to push for change.

Digital transformation also requires digital leadership and constant learning. The balance between soft skills, which are sometimes hard to develop, and hard skills, which are sometimes easy to buy, has become a necessary skill for digital leaders. A manager who excels in these two types of skills, hard and soft, understands better how technology trends, how to address people issues, and how to deal with legacy and transition challenges. They are better at negotiating deals with a variety of stakeholders and managing resistance.

A new and emerging best practice that has been adopted by many leading organizations is the creation of a new position called Chief Digital Transformation Officer (CDxO). The goal of this position is to centralize the process of change to create ownership at the top. Because digital transformation cuts across many layers of the organization and cannot be owned by a single function, the CDxO's role is to lead, centralize, and coordinate digital transformation efforts among the C-Suite and functional business leaders.

As a Digital Transformation Officer, your role is both tactical and strategic. As a tactical role, you need to define the processes and metrics that ensure that the transformation is on track. As a strategic role, you need to be able to identify the emerging technologies and the business models that will impact the delivery of products, as well as the customer experiences. You need to teach people how to embrace change and uncertainty and nurture in them the curiosity to learn and develop. A good Digital Transformation Officer simplifies the working environment by creating tools, policies, and procedures that support employees' transition. To be a true transformational catalyst, a Chief Digital Transformation needs to do the following:

- Help the organization develop structure, processes, and procedures that will serve as the engine of the transformation.
- Teach employees about the changing nature of the customer's needs, the economics of competition, the impact of exponential technologies, and the future of work.
- Create and develop knowledge to help leadership and employees see the unique way the organization performs in the market, and how the digital transformation fits into the value chain of the organization.
- Through negotiation, collaboration, and support, a Digital Transformation Officer helps other managers and key decision-makers understand technology trends and shifts and the added value the transformation brings.

TOOLBOX 5

THE NEW SKILLS OF THE NEW DIGITAL WORLD

A study conducted by Deloitte University Press in 2017 on the role of the CIO in driving digital transformation found that more than 67 percent of CIOs surveyed consider that digital disruption is the most critical aspect of the business that they must pay attention to. The study found that more companies are using digital to "do business in fundamentally new ways and different ways" rather than simply using digital to improve cost and operational efficiency. When it comes to the skills a CIO has to develop to succeed in today's digital world, the study found that experience in dealing with technology and knowledge about the type of technology that has an impact on the market should make them the leaders of choice for the transformation. When CIOs can combine their technology knowledge with business strategy, CIOs have better chances to leverage ecosystems and create talent and cultural models that challenge the status quo.

The study also identified a few skills that are a must for digital transformation. These skills are experimentation, risk-taking, and willingness to speak up. The study concludes with some key recommendations for organizations engaged in digital transformation. Following are some of the main recommendations:

- IT leaders are digital leaders, but they are not the only ones. While they hold resources and voice, according to the study, they cannot achieve their mission without business leadership skills. Adding business skills to their technology experience has the potential to make them true disruptors, especially when those business skills are combined with cross-functional collaboration, investment, and commitment to change, the study adds.
- Digital is not an initiative; it is a mindset. A common pitfall to digital transformation is to think of digital as a project or initiative. CIOs, and IT leaders, in general, should embrace a mindset that allows them to approach digital as a way of running the business. When IT leaders engage in the implementation of a new technology, their impact goes beyond the IT department.
- Customer centricity: IT leaders, who invest the extra mile to learn about the customer base of the organization and learn how to interact with customer data, provide better insights to decision-makers on customer pain points. IT leaders with a customer focus help C-suite leaders understand better the market in which they operate, the needs and wants of their customers, and how customers connect with the brand.
- Culture: Finally, the study concludes that for IT leaders to become true digital change agents, they need to nurture a culture of collaboration and teamwork.

For more information on this study visit the following source: Deloitte University Press (2017). Rewriting the Rules for the Digital Age: 2017 Deloitte Global Human Capital Trends.

Reskilling and Upskilling

When digital strategy gets broken down into a set of objectives, goals, tasks, and KPIs, coordination, communication, and collaboration become of paramount importance. A common digital transformation challenge is how to break through human behavior and bureaucratic barriers to communicate the essence of the transformations. Many leaders, absorbed by their KPIs, tend to focus on a single technology or the technology that serves them better, with a narrow KPI in mind, ignoring the fact that transformation requires a comprehensive approach and a new set of skills that have to be rebuilt. A LinkedIn data study, called the "great reshuffle," shows that the skill sets required for a job have changed by 25 percent since 2015 and that this number is expected to double by 2027. This is a critical challenge for business leaders. The skills required to do today's jobs are changing faster than we can adjust, and what is compounding this challenge is that the nature of jobs itself is changing as a result of the introduction of new cognitive tools. Generative AI, for instance, is expected to impact 40 percent of the UK job market, according to a KPMG report. This will require employees and individuals to seek new skills and new training that help them make the transition.

Furthermore, the remote work that was accelerated by pandemic effects added another layer of complexity by requiring skills, such as autonomy and independence, that are sometimes hard to find.

This is why leaders must invest in skills that are interdisciplinary and lateral by nature such as innovation and critical thinking.

Teach Critical Thinking Skills

The acceleration of change and the information overload to which employees are exposed requires an independent mind that allows the person to decide on their own. Employees who have high technical skills but lack critical thinking skills will find it very challenging to make decisions and solve complex problems that involve many variables on their own. Critical thinking is a skill employees learn and develop over time. It helps them conceptualize and synthesize data and information to conclude while looking at the issue from different angles. Making inferences is the most important dimension of critical thinking. Inference requires the ability to read data, analyze it, and make an assessment of the decision to be made. When a team member lacks this skill and has constantly to go to their manager for clarification, transformation stalls.

Creative thinking and innovativeness in approaching problems assume that the person has freed himself or herself from biases and mental traps. Cognitive traps, such as the anchoring trap, the status quo trap, and the sunk cost trap, are common mental traps that are hidden in our subconscious. They limit our ability to question assumptions and go beyond what we have been accustomed to. A common method of helping employees develop creative thinking is the six-hat exercise developed by Edward de Bono. This technique teaches learners how to approach a problem from different angles, how to question assumptions, and how to test new solutions. This exercise and many others are meant to help employees engaged in a transformation learn new lateral skills that they can apply to all sorts of problems they face.

Teach Innovation

In recent years, innovation acquired a new meaning and a new role in corporate strategy. Driven primarily by changes in demographics and digital technology, innovation has emerged as an imperative of corporate strategy. In a survey conducted by KPMG in 2021, more than 70 percent of CEOs consider innovation as critical to their ability to grow, compete, and earn new market share.

However, for innovation to bring value to the organization, innovation has to be managed as a system, so different employees and teams engaged in digital transformation have a common understanding of what innovation is and how it contributes to transforming their business.

A good starting point to design an ecosystem that teaches innovation and engages people with digital transformation is to start with ISO standard 56002: 2019, Innovation Management System. In this ISO standard, a viable innovation ecosystem requires the following five building blocks:

- **Strategy**: Design an innovation strategy that makes innovation the cornerstone of your digital transformation. When employees see the link between what they do, and the digital transformation, they engage in a meaningful way.
- **Communication**: A strategy that is not well communicated has a limited impact. So, design awareness and communication campaigns to teach employees and management about the role of innovation as a lever to transforming the business.

- **Incentives**: Establish appropriate incentives, including non-financial incentives, such as rewards and recognition. When employees see the link between their innovative ideas and the incentives they get, they will engage. Help them create linkages between innovative thinking and digital transformation so they can see the benefit of changing.
- **Tools and IP framework**: Create a framework to protect intellectual property to incentivize creative employees to engage with innovation and creativity. When innovative employees willing to transform and change the organization see that their intellectual product is not protected, they get discouraged from engaging in ideas and intellectual work. Employees are willing to take risks when they see that the company is willing to stand behind their IP work.
- **Teams and diversity**: Encourage the formation of diverse teams with a mix of people including different disciplines, personal attributes, and backgrounds to encourage cross-pollination that can challenge existing mental schemes and push people out of their comfort zone to change and transform. Reimagining the new business requires a mix of ideas.

Build a Customer-Centricity Capability

The disruption that has swept through the world and the change in demographics have made it clear that for organizations to stay relevant, they have to embrace a customer-centric philosophy in which the customer is put at the center of the transformation. Driven by AI-powered tools, customer centricity emerged as a force of change, and as a critical impetus to the digital transformation. The focus on the needs and wants of the customer, and the process by which the customer gets in touch with the brand, has forced business leaders to seek novel technologies that help them understand the economics of customer behavior.

An outside-in approach, in which the voice of the customer gains a seat at the table, serves now as the foundation of strategy formulation and strategy implementation. Efforts to individualize the customer experience have gained momentum, while innovation technologies, such as customer analytics, chatbots, and other conversational AI tools, have enabled organizations to learn how to anticipate customers' needs and how to proactively manage their experience. AI applications used in customer service have empowered companies to get closer to their customers to understand how customers think and make decisions. As well said by Steve Jobs, "Get closer than ever to your customers. So close that you tell them what they need well before they realize it themselves."

But how do you create a customer-centric philosophy in organizations, and how do you encourage employees to embrace it? What methodologies do you need to teach everyone in the organization to think in a customer-centric way?

Here are a few steps that you can take:

- **Engage your senior leadership** to invest in upgrading the current customer experience strategy to include AI-based applications that track customer behavior in real time. Take advantage of AI-based technologies to engage with the customer, understand their motives, how they get in touch with your brand, and what experience they are left with after using your product.
- **Conduct real-time market watch** to engage in dynamic retargeting, a strategy that allows you to target new demographics based on their previous contact with your brand, products, and offerings.
- **Use insights and aggregate data** from different touchpoints to understand when and how customers enter into contact with your offerings and what impression they are left with.
- **Engage with your customer** to co-create options that the customer wants, as well as address the pain points the customer feels.
- **Test and experiment** with your customer journey through a variety of ways to gauge its seamlessness. Feel the key moments the customer goes through when going through your customer journey. Figure 4.2 provides a five-step process for creating a customer-centric AI system.

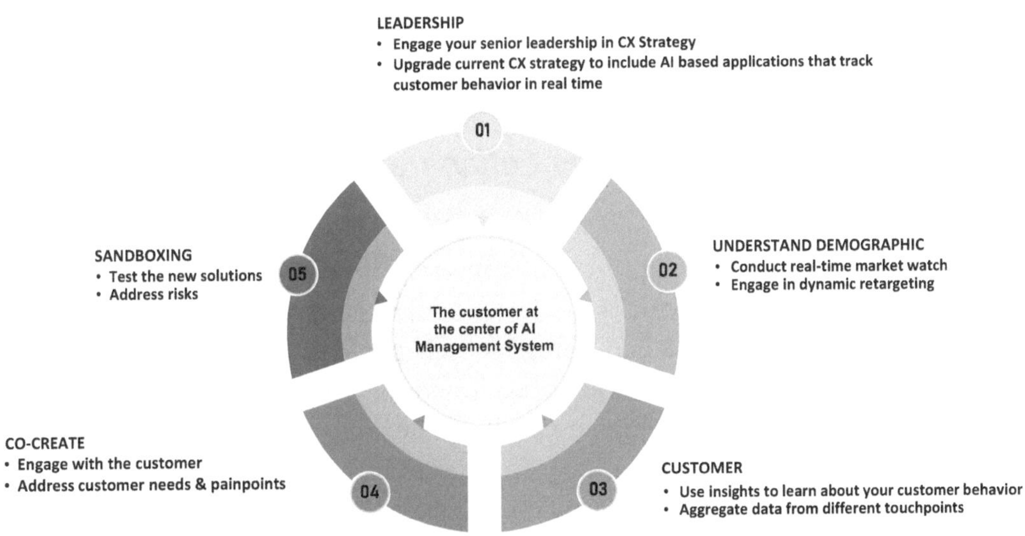

Figure 4.2 Build a customer-centricity capability.

Build an Enterprise Agility

Agility, as a concept, refers to the ability of the organization to respond swiftly to market changes, while still focused on the core. Relatively new to the literature on transformation, agility has been defined in many ways. Eric Abrahamson, for instance, defines it as the ability to achieve dynamic stability, in which decisions about strategy, focused on the core, are dynamic and take into consideration external market forces. Aaron Smith, a principal at Mackenzie, defines agility as the ability of the organization to be dynamic, nimble, flexible, and able to move fast to address the change.

In the age of disruption, adopting an agile framework to manage social, economic, cultural, and technological shocks is of paramount importance to the survival of the organization. The traditional way of focusing on stability and efficiency seems to be challenged by market volatility that requires speed and flexibility. A study conducted by McKinsey in 2020 found that the benefits leadership can reap from an agile approach are tremendous. Agility in organizations increases customer satisfaction by up to 50 percent. It also impacts positively employees' engagement and improves time to market, according to the same study.

But how do you create an agile organization? There are many methodologies you can adopt to engage the organization in agile implementation. A focus on a short strategy cycle and the design of a flexible structure are two examples. However, the two most critical dimensions of an agile framework are the creation of self-directed teams and the design of nimble processes that can be easily modified when things change. Let's review these two main ideas. Figure 4.3 provides a summary of the different steps that help you infuse agility into the entire enterprise.

Self-Directed Team to Manage Collaboration

Organizational structure is an important tool that helps a company deploy its strategy, allocate resources, and track progress. The typical organization design we are all familiar with is the functional approach, in which people are assigned to departments according to their professional background, and the matrix approach that combines both functional and divisional approaches. Both approaches have shown their limits in dealing with disruptions in general, and with digital transformation in particular. In a traditional structure, the bureaucracy of approval lines slows down the decision-making process, while forces that trigger change are fast and brusque.

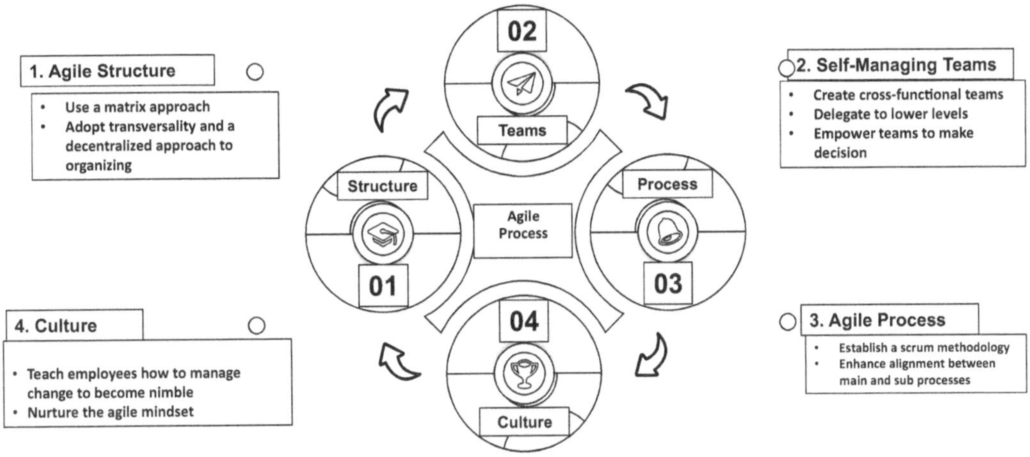

Figure 4.3 Building an enterprise agility: a four-step approach.

Creating self-governing teams that work through organizational hierarchy, speed up the decision-making process, and move things is a critical step in creating an agile organization. As well argued by Stanley McChrystal, the former commander of the Joint Special Operations Task Force in Iraq, the complexity of the world in which we live requires organizations to adopt a networked and decentralized approach. In his book *Team of Teams*, he outlines the need to rethink the traditional way of decision-making built around the concept of command and control.

- **Delegate to lower teams**: Hierarchy tends to create a bottleneck. When a manager has to approve all the output coming from below, this affects naturally the speed of the decision. Managers choose not to delegate for many reasons, including the lack of trust, lack of expertise at the lower level, or simply previous negative incidents. Team members who are on the ground tend to know more than the manager about the real issues they face. In his book *Team of Teams*, McChrystal argues that he often, as a team leader, was asked to approve decisions that people under his command knew of their context better than him. This slowed the speed of the action needed to be taken.
- **Encourage diversity**: While diversity may seem to be a barrier to fast decision-making, it is a good practice to have in teams. Diversity promotes divergent thinking and helps team members develop cognitive skills that help them understand the complexity that surrounds the context of the decision. Diversity is also important as it helps teams

avoid groupthink, a common pitfall in which people over time tend to develop similar thinking patterns that make it difficult for them to see the difference.

Agile Process: Review Your Decision-Making Process

To manage a digital transformation, you need procedures that are fast and nimble. You cannot decide on a digital transformation issue in the same way you decide on budget planning issue. Companies struggling with implementing processes struggle because either the process is too complex, that is, not intuitive, or it has not been well taught to employees. A process that is ill-designed, or not well-explained, creates bottlenecks for the workflow and adds more waste. People spend endless time in meetings asking basic questions about steps that are sometimes obvious. We often hear people saying that the process is getting in the way of transformation. This can be true only if the process is complex for people to digest and understand. Mature organizations invest tremendous time to create intuitive processes that are standardized so people can implement them seamlessly.

Generally speaking, two problems hinder the performance of a process:

- **Lack of governance**: In this situation, the process is not well mapped out or the person in charge is not identified correctly on the process map. People spend a lot of time figuring out where to go to execute simple tasks. Spend time mapping out different workflow procedures that relate to the digital transformation's programs and activities.
- **Lack of employees' skills**: The second obstacle that tends to hinder the performance of an agile process is the lack of skills and information. A sub-process can hinder the main process if employees lack the competence they need during the execution of that process. Here you need to spend time teaching people how to use the process and provide them with information, knowledge, and training to become nimble. Create a common language around that process so you can easily standardize it.

2 ARTIFICIAL INTELLIGENCE MANAGEMENT SYSTEM

How to Put in Place an AI Governance System

Introduction

In the first part of the book, I addressed critical issues that AI, and generative AI in particular, have brought to the world. We discussed global trends, the economic and social impact of AI on society, and some of the ethical and legal challenges that business leaders should address.

In the second part of this book, I will focus on the ISO/IEC Standard 42001 Artificial Management System subject of this book. Specifically, I will address the sections and clauses that address the implementation of the Artificial Intelligence Management System (AIMS) using examples and best practices from the industry. This part will follow the same high-level structure of the standard and will focus on the management clauses with some reference to the controls listed in Annex A of the standard.

DOI: 10.4324/9781003463979-6

This part will be organized into six chapters that follow the standard themes I chose to address. Particularly, I will address the following topics:

- Context of the organization (Clause 4): In this section, I will discuss the importance of context analysis. I will focus the discussion on why for AI, context is critical and what organizations need to do to analyze their context. My goal is to help you link your AIMS with what is happening in your organization's surroundings in terms of economic and business trends, regulation requirements, ethical expectations, and social issues that may impact the performance of your organization.
- Leadership (Clause 5): In this section, I will address the role of leadership in guiding the organization through the AIMS implementation process, as well as the roles and responsibilities of different management layers in facilitating the AIMS installment.
- Planning (Clause 6): In this section, I will explain how organizations should plan the process of putting together the AIMS and discuss the steps that managers need to take to conduct risk analysis. I will tie this section to the previous section on context analysis (Clause 4 of the standard) to highlight the need to connect the planning process with the context of the organization.
- Support (Clause 7): In this section, I will discuss the different actions leadership should engage in to prepare for the installment of the AIMS. I will focus particularly on the competence framework, and the communication strategy, as well as why the documentation is needed for audit and certification purposes.
- Operations (Clause 8): In this section, I will explain how organizations should manage, mitigate, and treat AI risks. I will also go beyond ISO/IEC 42001 to provide examples from other frameworks such as NIST's framework to help you develop a more comprehensive approach to risk analysis and risk treatment.
- Performance Evaluation (Clause 9): In this section, I will discuss and explain how organizations design a performance evaluation process that allows you to track, monitor, measure, and evaluate the overall AIMS performance. I will provide metrics examples and KPIs used in the industry as well as some best practices on how to manage AIMS performance evaluation.
- Improvement (Clause 10): In this final section, I will explain the steps organizations should engage in to collect feedback about the

performance of the AIMS, and how to use this feedback to improve the overall performance of AIMS using a PDCA approach.

- Scope and Normative Reference: (Clauses 1 and 2): These sections of the standard are part of all ISO management system standards and will not be addressed in this book. They are easy to read and understand.
- Terms and Definitions (Clause 3): I will not address this section. However, it is a critical section to read and understand. It provides definitions of the terms and the concepts adopted in the standard. Reading it helps your team and those involved in the AIMS implementation develop a common language and a better understanding of the terminology used in the standard. I suggest that you take a look at it with your team to minimize interpretation issues. Standards implementation goes much smoother when everyone has the same reading of the text (see Figure P.1).

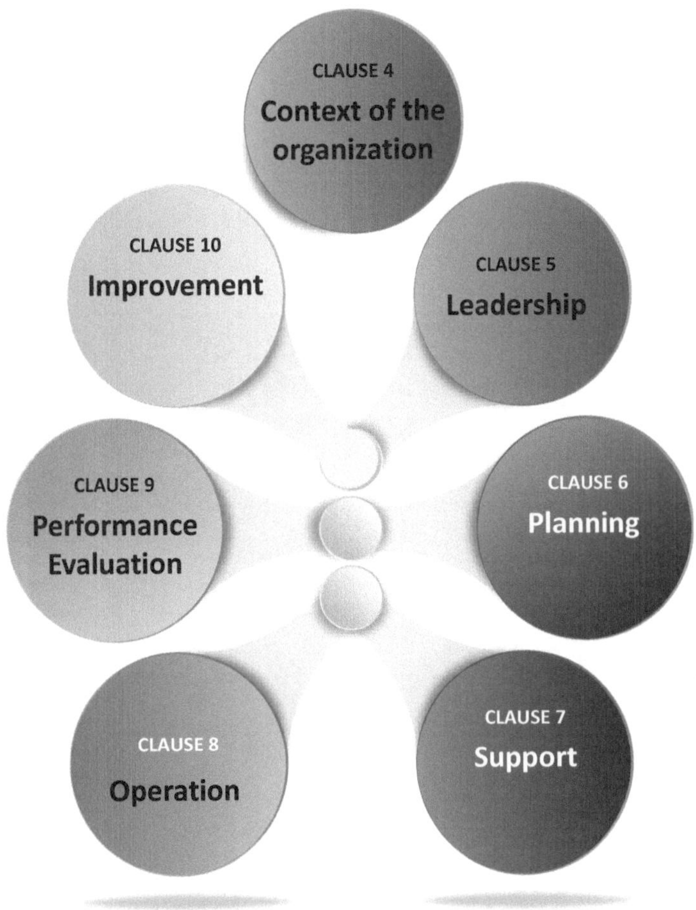

Figure P.1 Organization of the book.

Chapter 5

Clause 4: Context of the Organization

PURPOSE

The purpose of this chapter is to discuss Clause 4 of the ISO standard. The chapter explains why the analysis of the organizational context is critical to setting up an AIMS and discusses the process by which an organizational context is conducted.

OVERVIEW

A requirement of the ISO/IEC 42001 Standard is to analyze the context of the organization. This chapter explains the process by which you analyze the context of the organization as well as why it is important to run the analysis. The chapter delves into the economic dimension of AI and legal compliance issues, as well as the ethical framework that you need to consider when establishing the AI management system. A focus on the EU AI Act and the US regulatory landscape will help you understand the global AI legislative trends and the regulatory shifting landscape.

DOI: 10.4324/9781003463979-7

BENEFITS

At the end of this chapter, the reader will be able to:

- Understand why context analysis is critical to setting up the foundation of AIMS.
- Learn about the risks and compliance and how laws and regulations apply to AIMS.
- Develop an ethical framework for responsible and trustworthy AI.
- Learn how to conduct an internal and external organizational analysis.

Introduction: Why Context Analysis Is Crucial to AI Management System?

Numerous reasons make you invest time, effort, and resources to study, analyze, and understand the context of your organization. Analyzing the context of your organization helps you first understand the bigger picture of the conversation on AI, and how your organization fits into that conversation. Context analysis provides you with a comprehensive view of the historical background, as well as the social and political factors that drive the discussion about AI. Context analysis is crucial for legal purposes as it helps you learn about how laws, regulations, and standards evolve in different countries, how these laws and regulations may affect your products and services, and your overall AIMS performance.

Furthermore, context analysis helps you identify key stakeholders that may be affected by your AIMS. When you analyze the context of your organization, you have a better perspective on how your AIMS affects people and customers, which in turn should provide you with your opportunities to improve the effectiveness of your decision-making process. Studying the context of the organization helps you also identify the links that exist between your products and services, your AIMS, and different market players. Learning about who is doing what in the AI sphere provides you with the opportunity to identify connections that link your products and services with other actors and players, so you are better informed about trends, emerging issues, and risks.

When looking at the context of the organization, the ISO/IEC 42001 Standard requires that the organization analyze and evaluate internal and external factors that may affect the performance of the AIMS. In the context analysis, you need to look at a variety of factors that may affect the performance of your AIMS and the performance of your overall organization.

First, you need to identify stakeholders (interested parties in the language of the Standard), such as providers, developers, users, producers of AI systems and applications, partners, AI subjects, and authorities. These stakeholders are defined by another ISO/IEC Standard (ISO/IEC 22989:2022 AI concepts and terminology) and are an important component of the analysis and a requirement of the standard. ISO/IEC 42001 also requires you to look at the ethics of your organization, the value system, and the management practices that govern the functioning of your organization. Putting together a new management system for a highly complex issue such as AI requires a fundamental mindset shift that is not easy to do.

Furthermore, and because ISO/IEC 42001 is a certifiable standard, a good context analysis is also crucial to the auditing and certification process. Conducting the analysis, and documenting this process correctly, gives you a better chance to pass the certification process. It helps the auditor go beyond technical analysis to understand your philosophical approach to AI and read better your intentions about how you deal with powerful and transformational technology such as AI. It shows the auditor your ability to think critically about transformational issues and how you anticipate future risks and challenges that may impact the performance of your AIMS, and the overall organization.

TOOLBOX 6

WHY UNDERSTANDING THE CONTEXT OF THE ORGANIZATION IS CRUCIAL TO A GOOD IMPLEMENTATION OF YOUR AIMS

Analyzing the context of the organization helps you understand four key elements: Understanding the big picture, identifying key stakeholders, learning about links and interconnectedness, and improving agility.

The Bigger Picture	*Identify Stakeholders*	*Identify Links*	*Improve Agility*
■ Understand the political, social, cultural, and economic forces behind the AI evolution. ■ Learn how society is shaping the discussion about AI, and how AI is shaping people's perception about your products and services. ■ Understand and identify the actors and the issues driving the AI conversation.	■ Learn about the different stakeholders that will be affected by your AIMS. ■ Understand how and in what way other players and actors may affect your AIMS. ■ Identify opportunities to collaborate or compete with stakeholders for a better AIMS implementation.	■ Learn about who is doing what in the AI sphere and how you can connect with them to improve your AIMS. ■ Identify opportunities to connect or compete with different AI players. ■ Learn about how to create coalitions to push and protect your interests.	■ Understand the changing environment of AI and how this may affect your AIMS. ■ Understand how to manage change better. ■ Learn about mechanisms that help you adapt better to the context. ■ Learn how to improve your responsiveness and anticipation strategies strategy.

What to Include in the Context Analysis

To help you understand the requirements of the Standard on how to analyze your organizational context, I will explain three critical issues:

- Competitive landscape and stakeholders' analysis
- Legal context, laws, and regulations
- Ethical AI and responsible AI

Competitive Landscape and Stakeholders' Analysis

When looking at the competitive landscape and stakeholders' analysis, your first step is to understand the technology trends and the disruption AI technology is having on your company and the overall competition. Conducting an industry analysis to understand how AI impacts your market and your industry is a critical step to learning about the organizational context. Depending on the sector in which you operate, an industry analysis helps you ask the following three critical questions.

- How is AI reshaping my industry and to what extent the AI trend is disrupting the overall competitive landscape?
- How is AI impacting key market segments and channels in my industry, and how is the competition responding to this trend?
- How is AI forcing the repositioning of current players while bringing new entrants that may be of threat to my existing products and services?

Many frameworks may help you address these questions, but a good well-known market analysis framework is Porter's Five Forces model. In this model, you analyze five elements:

- The competition and how AI is impacting the industry in general.
- The potential threat of new entrants from AI startups and from companies that have adopted AI technologies.
- The power of third-party AI suppliers, such as developers, designers, and platform providers.
- Customers' choices and customer trends.
- Substitute products from AI that may displace your existing products and services.

Once you conduct the market analysis, you turn to the analysis of internal organizational capabilities. In the internal analysis, your goal is to understand to what extent the AI-induced change in the market affects the internal functioning of your organization. This will allow you to think critically about how to prepare your organization for the change and transformation AIMS will have on your existing capabilities.

Here are the five guiding questions that you need to ask.

- What resources, financial, human, physical, and technological, do I need to invest in to make my AIMS a viable one?
- What stock of intangible resources do I need to build or invest in to make my AIMS a better-performing system? Intangible resources such

as knowledge, trust, innovation, reputation, ideas management system, R&D, and brand.
- What organizational capabilities do I need to develop or enhance to make my AIMS a successful one? An organization's capability is the capacity that helps you deploy resources in a unique way that is particular to you. Capabilities tend to emerge over time, and they are a combination of many things including knowledge, skills, talents, network, and expertise.

Finally, the standard requires a stakeholder analysis. In this analysis, you need to identify the different types of stakeholders involved with your AIMS as well as their needs. While most organizations should have already identified their main stakeholders, in the case of AI, and depending on your sector of activity, a new list of stakeholders is a must. This is a requirement of the standard that must be documented. The standard lists six types of stakeholders that are well defined in the ISO/IEC 42001 and the ISO/IEC 22989:2022 AI concepts and terminology. Figure 5.1 provides a 15-step process that helps you run an organizational analysis.

TOOLBOX 7

HOW TO PREPARE YOUR ORGANIZATION FOR THE ASSESSMENT, AUDIT, AND CERTIFICATION PROCESS

To prepare for the assessment, audit, and certification process of your organization, start with Section 4. Section 4 of ISO/IEC 42001 requires the following actions:

- Analyze the context of the organization to understand how different internal and external forces shape the development and the functioning of AI systems, and how AI systems impact the overall performance of the organization.
- Study different stakeholders' relationships with the organization to identify their needs, expectations, and their requirements to create an integrated AI management system.
- Define and document the scope of the AIMS, its boundaries, where it starts and where it ends, as well as its relationship to other internal

management systems. This should be defined in a document that will be presented during the auditing process.

- Develop a holistic approach to AI by putting together a comprehensive AIMS that includes all components of ISO harmonized management systems. The components of harmonized management systems are leadership, planning, support, operations, performance, evaluation, improvement, controls, and objectives.

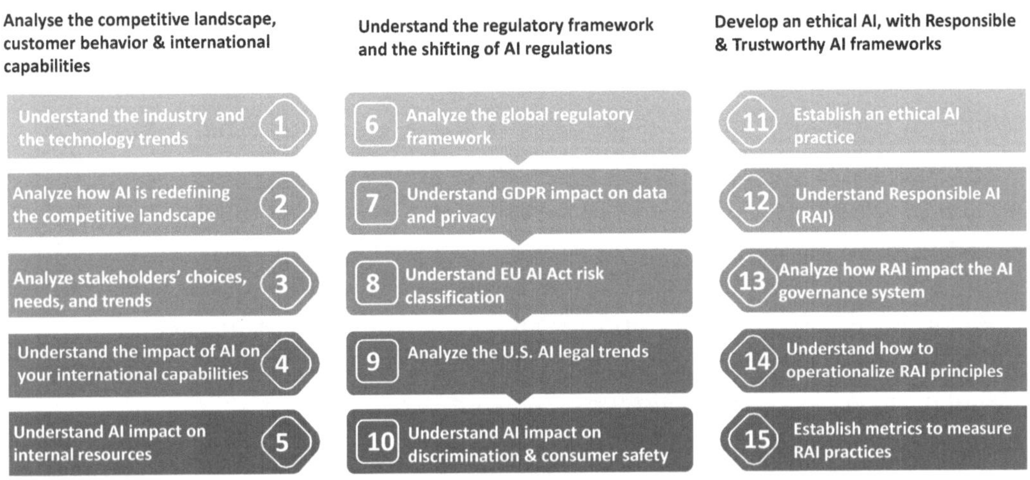

Figure 5.1 A 15-step process to conducting an AI context analysis.

Legal Context Analysis: Laws and Regulations

The second component of your organizational context analysis is legal. Because of its powerful and transformative nature, AI is poised to impact every aspect of our lives. As a general-purpose technology, AI will have a broad impact on people's lives, so understanding the liability and the risk of this technology, as well as the legal exposure of those who provide AI products and services is critical. This is why understanding the legal context of AI is an important step to the viable functioning of your AIMS. While AI-specific regulations are changing by the day, as many countries and jurisdictions are catching up to the speed of technology advances, it is nevertheless helpful to pay attention to some of the legal frameworks and principles that have been already adopted. As of the writing of this book, the EU General Data Protection Regulation, the EU AI Act, the US AI Bill of

Rights, the OECD principles, the UNESCO AI principles, and the NYC Local Law 144 of 2021, which prohibits discrimination in Automated Employment Decision Tools (AEDT), provide a good indication on the future of AI regulations, the global legal trends, and regulators intention in different jurisdictions. They serve as a reference point to the future of the AI regulatory framework.

Following is an overview of three legal frameworks that have been either adopted or are being adopted as of the writing of this book: The General Data Protection Regulation, the EU AI Act, and some of the US AI regulatory initiatives launched by federal agencies.

The General Data Protection Regulation

GDPR is a legislation that was adopted in the EU in 2018. The goal of the GDPR was to unify different European legal frameworks that address privacy and provide EU citizens with legal means to protect their personal information. It was also meant to provide EU citizens with the means to file a complaint against violation of these rights as it provides many key definitions and establishes liability and penalties for violation of the act.

While the GDPR does not address AI per se, it defines many concepts that relate to AI systems. Following are some of the issues that GDPR addresses:

- **Personal data**: Information that can identify, directly or indirectly, individuals.
- **Data processing**: The process of manipulating data during all the stages of data usage, from collection to recording, structuring, and storing the data.
- **Data subject**: This refers to customers, employees, partners, and all users of data and AI systems.
- And finally, **data controller**, which refers to the person who has decision authority on where and how the data is used. This is a critical definition because it puts the liability directly on the data processing governing body.

Furthermore, the GDPR establishes seven principles to protect and account for data processing. All these principles could have a direct link to ISO/IEC 42001. These principles are:

- Processing data must be used for lawful purposes. It must be fair and transparent to users.

- Data must be processed according to the purpose for which it has been collected (purpose limitation principle).
- Organizations collecting data should limit data collection to only the data they need and avoid casting wider nets that give them more data than what they need (this is the data minimization principle).
- Data must be accurate and kept updated.
- Organizations should store only the data they need to avoid storing data that they don't need (the storage limitation principle).
- When processing data, data must be protected (by encryption for instance) to respect the integrity and confidentiality of the data.
- Organizations processing data are liable for compliance with GDPR principles.

All these legal stipulations directly affect your AIMS. The analysis of the context should spell out the process by which your organization intends to address these legal requirements. Keep in mind also that these GDPR legal requirements are similar to many of the standard clauses that relate to data processing. Addressing them at the same time and documenting this process would be the right thing to do as it saves you time and helps you get ahead of the game.

The ISO/IEC 42001 requirements that relate to data processing and that are similar to the GDPR legal requirements are listed as normative controls in Annex A. These controls are B.4.3, B.7.2, B.7.3, B.7.4, B.7.5, and B.7.6.

TOOLBOX 8

DATA MANAGEMENT CONTROLS

The standard ISO/IEC 42001 provides a set of controls that organizations seeking certification need to have. These controls are listed in Annex A of the Standard. Organizations can implement these specific controls listed in the standard or develop their equivalent controls. Concerning data management, listed under Clause B7 Data Management, the standard requires five actions that have to do with the process of collecting data, how to make sure that you have good quality data, and the overall processing of data that could enhance the overall performance of the AIMS.

The EU AI Act

In April 2024, the European Parliament approved the EU AI Act, a landmark legislation that has been in negotiation for some time. This is the first AI legislation in the world that aims at regulating AI, and it is the first legislation that specifically addresses generative AI systems.

There are a few key ideas that characterize this legislation.

- First, the act focuses on Machine Learning systems and intelligent systems that provide recommendations, predictions, and statistical models that may impact the decision-making process, thus excluding traditional software and applications. By doing so, the act brings into focus the discussion on AI automated systems, intelligent systems, and foundation models. Examples of intelligent systems subject to the act are machine learning systems, including supervised, unsupervised, and reinforcement learning, logic and knowledge-based approaches, statistical, and Bayesian approaches, search and optimization methods, and large language models.
- Second, the act makes an obligation for companies developing foundation models to assess and mitigate risks to health, safety, environment, and human rights. Foundation models providers must register their models in the EU database before releasing them to the public. They must also comply with the transparency rules, such as disclosing the copyrighted materials and data used to train foundation models.
- Third, the act bans AI for biometric surveillance, emotion recognition, and predictive policing. These AI systems are deemed so dangerous that they are simply made illegal.
- Finally, the act adopts a risk-based approach in which it identifies four levels of risks of AI systems. Unacceptable risk, high risks, limited risks, and minimal or no risk. Each category of risks has a specific set of rules and controls as explained here. Figure 5.2 explains the EU AI Act risk classification system.

Unacceptable Risk

In this category of risks, algorithmic systems that use social scoring, facial recognition, and subliminal techniques that may cause harm are simply prohibited. Examples of such systems are:

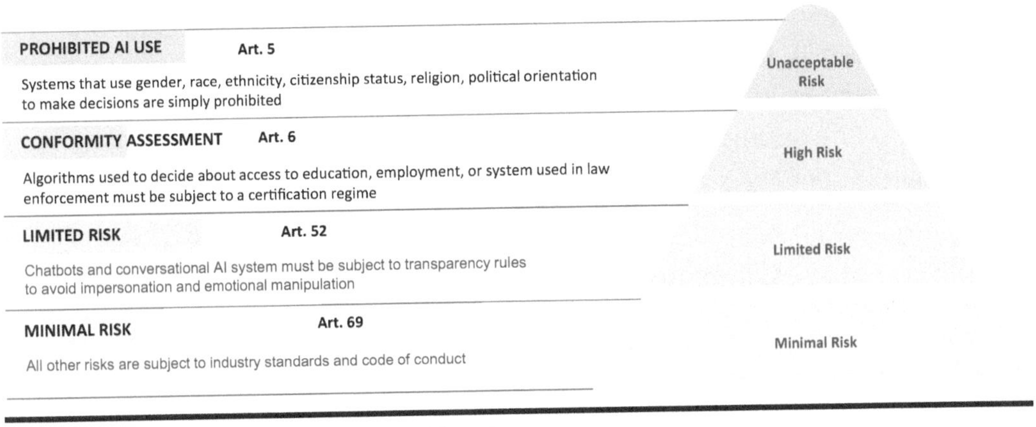

Figure 5.2 The EU AI act risk classification.

- Systems that use gender, race, ethnicity, citizenship status, religion, political orientation to make decisions.
- Systems that identify location, past criminal behavior, and profiling, such as in predictive policing.
- Real-time remote biometric identification systems such as facial recognition.

High Risk

In this category, AI management systems are subject to a regime of certification, controls, and conformity assessment to test their compliance with specific and existing laws and regulations, and to make sure that their output does not create harm. All AI management systems that relate to critical infrastructure, biometrics, education, training, employment, and essential public services, like welfare benefits, law enforcement, migration, and border control systems, and the administration of justice, are considered high-risk AI management systems and therefore subject to a certification process. From an organizational perspective, most of the high-risk management systems fall under the functions of HR, accounting, finance, customer service, IT support, and legal.

Limited Risk

In the limited risk category, the EU Act identifies systems that relate to chatbots and deepfakes. Deployers of these types of systems must comply with transparency requirements. They must disclose that the interaction with the chatbot is an interaction with a machine, not with a human, to avoid harm related to deception and risk of impersonation. Providers of these systems must ensure that users are informed that they are dealing with a chatbot, not with a human.

Low Risk

In the low-risk category, AI systems are not subject to the EU AI ACT. They are left to the ethical frameworks and other norms and values that regulate the overall industry.

Finally, it is important to mention that AI applications do not operate in a legal vacuum. In many instances, they are regulated directly or indirectly by the already existing laws and regulations in many countries. Legislation that pertains to antidiscrimination, workplace safety and health, consumer protection, and credit and finance can, and may, be applied to decisions that are made by AI systems, especially when there is evidence of harm occurring as a result of the automated decision. So, it is critical that the context analysis takes a broader perspective and identifies possible legal implications of different risks, including the unknown ones that may be the result of the rapid rise of generative AI and Large Language Models.

TOOLBOX 9

REGULATING AI: A COMPLEX CHALLENGE

The challenge of regulating artificial intelligence (AI) has garnered significant attention. In a piece published by the Brookings Institute, author Tom Wheeler suggests an interesting perspective. He argues that AI poses unique regulatory challenges due to its distinct nature as a technology. The rapid pace at which AI evolves presents a big challenge for the US government.

Wheeler identifies four primary challenges:

1. **Velocity**: Unlike previous technologies, AI exhibits remarkable speed in both development and adoption. OpenAI, for instance,

gained 100 million users within five days, while other generative AI applications, such as Mid Journey, became a global sensation in a short period. The government's structure, rooted in assumptions from the industrial era, is simply ill-equipped to cope with such rapid technological progress.

2. **What to regulate**: AI is a highly complex and multifaceted technology. Containing it in one piece of legislation simply will not work. As a general-purpose technology, AI permeates many aspects of our lives, so a one-size-fits-all regulatory approach is simply inadequate, according to Wheeler.
3. **Who regulates?** The structure and philosophy of the US government were designed during the Industrial Revolution when innovation and technological advancements pace was slower. At that time, the government had sufficient time to study, investigate, and conduct hearings to regulate innovations. However, the lack of agility within the current governmental framework presents a major barrier to keeping up with AI technological advances.
4. **How to regulate**: The notion of licensing AI, the notion that has been advocated by many business leaders such as Sam Altman, the founder of OpenAI, poses its own set of challenges. Wheeler argues that the licensing system, like in the case of the FDA and FCC, can impact the competitive landscape. The initial license holders tend to dominate the market and subsequently advocate for new rules that may impede the entry of new competitors.

Source: Wheeler, T. (2023) The three challenges of AI regulation. The Brookings Institution. Retrieved from https://www.brookings.edu/articles/the-three-challenges-of-ai-regulation/

The US AI Regulatory Landscape

In the United States, the regulatory framework of AI and autonomous systems is still fluid and evolving. Many federal agencies, states, and other local government agencies have issued regulations, directives, and guidelines to address the issues of discrimination and bias in automated systems. According to a review by author Ravit Dotan (2023), and as of the writing of this book, the following federal agencies, states, and local government entities have issued the following directives and regulations:

- **The joint statement on the enforcement effort against discrimination and bias in automated systems** issued by the Civil Rights Division of the US DOJ, the Consumer Financial Protection Bureau, the Federal Trade Commission, and the Equal Employment Opportunity Commission. In this statement issued in April 2023, software tools and new intelligent systems, such as AI used to organize workflow, are subject to the current legislation that protects civil liberties and individual rights, particularly when discrimination is the result of how organizations use data and datasets, how their model is trained, and how the entire a system is designed.
- **Assessing adverse impact in software, algorithms, and artificial intelligence used in employment selection procedures under Title VII of the Civil Rights Act of 1964**: A federal guideline was issued in April 2023 by the Equal Employment Opportunity Commission to clarify the use of automated tools in HR selection procedures. The EEOC reiterates the idea that automated tools are subject to Title VII of the Civil Rights Act of 1964. Organizations may be responsible for disparate treatment and adverse impact, two core concepts to Title VII, during the HR selection decision-making process. Employers, according to this guideline, may be even liable for third-party tools, such as employer's agents and vendors of AI tools and automated systems.
- **Equal Credit Opportunity Agency (ECOA)**: The ECOA introduced a change in its regulation to make it illegal for the use of biased algorithms. Discrimination in automated systems that are based on race, color, religion, national origin, sex, marital status, age, or public assistance benefits is prohibited.
- **Fair Credit Reporting Act**: The Federal Trade Commission took enforcement actions and issued directives to emphasize the need for companies to pay attention to discrimination in automated systems in employment, housing, credit, insurance, or other benefit.
- **California Consumer Privacy Act**: This act provides more protection to consumers on how companies use their data. According to this Act, consumers have the right to their data during all the data collection stages, including the right for consumers to opt out.

- In the financial service, the Office of the Controller of the Currency (OCC), the Consumer Financial Protection Bureau (CFPB), the Federal Trade Commission (FTC), the Financial Industry Regulatory Authority (FINRA), and the Commodity Future Trading Commission (CFTC) all issued directives in which discrimination and bias in automated systems are directly or indirectly addressed.
- Finally, in the medical field, the FDA released guidance recommending that certain predictive devices that are powered by AI applications should be regulated as medical devices.
- **At the state level**, and according to some estimates, 14 states enabled or adopted legislation on AI, bringing this to a total of 26 bills as of 2023. The Bills cover many aspects of AI risk, from harm to consumers to facial recognition and hiring discrimination. In the insurance industry, and the state of Colorado, according to researcher Shea-Brown, external data as well as algorithms and predictive models that discriminate against a protected class are prohibited.
- **New York City Local Law 144** Perhaps the most pronounced legal framework in the United States that addresses directly bias and discrimination in AI is the New York City Local Law 144 issued by the NYC Department of Consumer and Worker Protection in 2020. To this day, this is the most aggressive AI regulation framework. It prohibits discrimination in Automated Employment Decision Tools and sets high penalties for violations. In this law, AEDT is defined as "any computational process, derived from machine learning, statistical modeling, data analytics, or artificial intelligence" that ranks and recommends applicants for an employment position. The law requires, among other things, that companies adjust selection tools to link them directly to essential job functions, mandating algorithm audits. When using a third-party assessment tool, the law requires that companies must audit the assessment tool and choose the least discriminatory assessment method available.
- **The US Algorithmic Accounting Act of 2022** in April 2022, a group of senators introduced the Algorithmic Accountability Act. The legislation would require organizations to conduct mandatory impact assessments of their automated systems. This act, which is an updated version of the Algorithmic Accounting Act of 2019, defines automated decision systems (ADS) as software or processes that use computation to produce outputs that are used to make decisions. ADS that impact education, employment, financial services, housing, and other services

critical to consumers' choice fall under the scope of the Act. The Act applies to organizations that have more than $50 million in average annual gross receipts. The Federal Trade Commission will be tasked to implement the Act. The Act was rejected by the Congress, and it never made it to become a law.

Finally, it is important to mention that at the federal level, the White House published the AI Initiative in 2019. This initiative aims to promote innovation, funding, and research in the AI field, and addresses issues that relate to privacy, civil liberty, and other national values. In 2022, the White House also published a blueprint for an AI Bill of Rights, in which the White House outlines many of the principles that promote trustworthiness and responsible AI.

The Executive Order on Safe, Secure, and Trustworthy Artificial Intelligence

In October 2023, President Biden issued a sweeping Executive Order that regulates AI. The goal of the Executive order is to protect American consumers; ensure that AI systems are safe, secure, and trustworthy; and help the United States build the capacity needed for US global leadership. The Executive Order also charges many federal agencies with regulating AI through standards and continuous monitoring of AI risks.

The Executive Order mandates the following:

- Organizations developing powerful foundation models that may affect national security, national economic security, or national health security must share with the US government the results of their safety tests and other critical information.
- The National Institute of Standards and Technology (NIST) is tasked with developing rigorous standards to ensure that AI systems are safe, secure, and trustworthy to be applied to protect critical infrastructure.
- Protect Americans' privacy through federal support and privacy-preserving research.
- Address the issue of algorithmic discrimination through training and best practices sharing.
- Support the responsible use of AI in healthcare and education.
- And finally work with other nations to expand bilateral multilateral and multistakeholder engagements to collaborate on AI.

Ethical AI: Responsible and Trustworthy AI

Because the legal standard may or may not apply to a specific situation, and because of the speed at which AI technology has been moving, ethics and ethical reasoning should be also considered during the context analysis. The goal is to show certifying bodies that the organization is aware of the risk, the ethics, and the moral challenges AI and automated systems are posing to society. In situations where there is a lack of legal clarity and where the use case is a case of gray area, ethical frameworks can be of great help to decision-makers as they provide guidance on what to do and not to do. The ethical reasoning is also captured by the spirit of the ISO/IEC 42001, so it will be helpful during the certification process to show that the AI management system are guided by ethical reasoning and subscribe to the objectives advocated by the Standard. Figure 5.3 provides a framework for responsible AI.

There are many ethical frameworks that organizations can implement. Some of these frameworks have been published by ISO that you can find in the following two standards:

- ISO/IEC TR 24028:2020 Overview of trustworthiness in artificial intelligence standard
- ISO/IEC TR 24368:2022 Artificial intelligence, an overview of ethical and societal concerns.

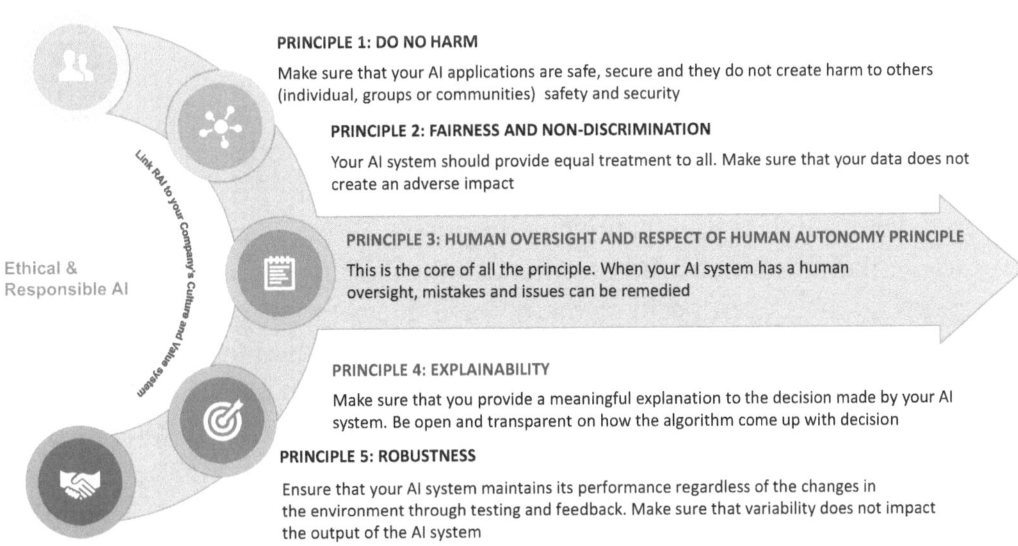

Figure 5.3 Ethical AI: Responsible and trustworthy AI principles.

Other frameworks and principles have been also published by international organizations such as the European Commission's High-Level Expert Group, UNESCO, and OECD. Technology and consulting companies, such as Microsoft, IBM, PWC, Deloitte, BCG, and many others, have also published sound frameworks. These frameworks are an excellent first step to establishing a RAI practice. They all agree on the need for senior management to engage in building a set of ethical principles that guide the organization during the design and implementation of AI applications.

However, the ethical principles that seem to get a wider consensus among different stakeholders are the following:

- Principle of do no harm
- Principle of fairness and non-discrimination
- Principle of human oversight (Human-in-the Loop)
- Principle of explainability
- Principle of robustness

These principles capture well the essence of fundamental human rights and should be considered as an ethical imperative as they help you build a sustainable and trustworthy AI system. They also can help during the certification purposes, as they align with the objectives of the ISO/IEC 42001 (Annex C Informative), as well as the spirit of many of the Standard clauses. Following is an overview of these principles and some suggestions on how they should be operationalized.

Do No Harm Principle

In philosophy, do no harm principle is a central and foundational principle. According to the liberal political philosophy, humans are free to act in any way they intend unless their actions harm others. When their actions may be a source of harm to others, then restrictions on their freedom may apply. For AI technology, harm can occur in many ways. It can occur in the way AI technology is used for surveillance purposes or social manipulation when individuals are coerced directly or indirectly to do things against their will. Harm can also result when biased decisions are made by the AI systems or during data processing. In extreme cases, harm can also occur in automated weapons that are fueled by automated algorithms. The prevention of harm in AI also requires that the AI systems be safe and secure and that they don't affect the moral and physical integrity and well-being of the individual.

In the context of AI, author Smuha (2021) distinguishes between three types of harm: the harm that affects the individual, like in the case of facial recognition that may negatively affect people of color; the collective harm, which may affect a group of people that share similar traits; and the social harm which, according to the author, goes beyond the individual and the group to affect the way a society functions. Examples of generative AI applications, such as the ones that produce disinformation at scale, may affect the social fabric of society, with harmful consequences on its political institutions and democratic process.

ISO/IEC 42001 requires that you define harm. Clause B.5.5, under the title "Addressing the Societal Impact of AI Systems," the standard requires that you assess and document the potential physical harm of your AI systems on users. You should also identify processes and tools that help you prevent harm and take corrective actions when potential harm is identified. Human dignity and the physical and mental health of users should not be compromised by AI technology.

TOOLBOX 10

HOW DO WE ASSESS THE DO NO HARM PRINCIPLE? ASK THE FOLLOWING QUESTIONS:

- Do our AI systems identify harm?
- Do we have a list of possible harm that can result from the use of our application, products, and services?
- Do we engage with stakeholders to identify the impact of our AI systems and identify harms from the user's perspective?
- What laws, regulations, ethical frameworks, and standards have we adopted to define harm in AI systems, applications, and products?
- Have we conducted an impact assessment of AI systems to identify possible harm?

The Principle of Fairness and Non-Discrimination

Fairness means providing an equal treatment. The issue of fairness has been debated in philosophy, law, and politics for some time, and now it has made its way to artificial intelligence and autonomous systems. The EU AI Act defines fairness as "a commitment to ensuring equal and just distribution of both benefits and costs and ensuring that individuals and groups are free from unfair bias, discrimination, and stigmatization." Simply said, fairness refers to the equal treatment of individuals when it comes to access to goods and services provided by the organization.

But how can this concept, which seems to be so interwoven with human nature, be translated into the context of Artificial Intelligence? How can a machine be fair or unfair? And how can we make sure that machines, which are supposed to be objective and neutral, act and produce fair results?

In Artificial Intelligence, a new field called algorithmic fairness emerged in recent years. The focus of this field is the study of machine learning models and algorithmic decisions that may create bias and adverse impacts on the way they process data and make predictions. Examples of algorithm bias have been widely discussed in the financial sector, where loan applicants may receive different treatments by the algorithms, and in the criminal justice system, where prediction has not been fair to a particular ethnic group.

In studying algorithm unfairness, it is critical to point out the fact that algorithm models by themselves are not the source of the bias. It is the data we feed into the model and the way we train the model that can be the source of discrimination. Data that we feed into the model when it is poisoned intentionally and unintentionally perpetuates discrimination. Machine learning models that are trained on biased data, or trained on non-representative samples, will reproduce discrimination.

To remedy this situation, researchers have suggested many methods and techniques. Methods such as equal odds positive, predictive parity, and counterfactual fairness are some of the technical suggestions that can alleviate the bias in predictive models.

TOOLBOX 11

HOW DO WE DE-BIAS ALGORITHMS?

The Center for Equity, Gender, and Leadership at Berkeley Haas Business School put together a playbook to help business leaders understand the issues of fairness and algorithm bias.

To help your organization de-bias algorithm, the center suggests three actions:

- **Diversify teams in charge of AI applications**: For the teams putting together AI systems, diversity seems to be a critical factor in helping developers build an AI system free of discrimination. Diversity of background, experience, and perspective helps teams avoid blind spots, groupthink, and many other cognitive biases that are the result of people being unable to question what seems to be obvious to them. More importantly, the center suggests the concept of interdisciplinarity in teams building AI applications. Organizations should engage individuals in social studies humanities and other fields that may not be the forte of typical AI engineers. Finally, providing teams with the necessary training on issues of fairness, bias, adverse impact, and disparate treatment, so AI developers and builders are aware of the issues.
- **Design AI systems with the Responsible Artificial Intelligence (RAI) Principles**: From the start, when you design an AI system include RAI principles. This can take many forms. Audit of the datasets used, documentation of the source and the origin of the data, and assessment of the representativeness of the groups and samples used in the data. The goal is to infuse RAI principles into AI applications during the design stage. The center suggests that teams ask the following questions:
  1. How do teams in charge of AI systems assess the quality of the data?
  2. How do we make sure that existing datasets are not used for other ML functions?
  3. Are AI systems audited internally and externally to make sure that bias is identified?
- **Governance and leadership**: Without governance and leadership involvement, none of the previous procedures work. When senior management and governance bodies are engaged in these principles, employees down the chain of command working on the actual development of AI applications are aware of the stakes, the harm, the risks, and the rewards that are attached to RAI principles.

Human Oversight and Respect of Human Autonomy Principle

The fundamental idea of this principle is that AI technology should not be in control of human will and actions. The oversight, governance, and control should be always in the hands of the humans. AI technology, from this perspective, regardless of how autonomous it is, should be always looked at as an enabler that empowers humans and complements their work. From this perspective, and according to the EU Commission, respect for the human autonomy principle means that a technology should not be built to control, deceive, or manipulate a human will. A human-centric approach to designing AI products in which the focus is on enabling humans rather than controlling them should be adhered to during all steps of designing and implementing the technology.

Another concept related to oversight is the concept of the Human-in-the-Loop principle well known in machine learning programs. Particularly for high-risk decisions, where human needs to provide the input, interact with the system, and review the output, the human-in-the-loop principle provides a critical safeguard for the training, the simulation, and the initial deployment of AI systems.

TOOLBOX 12

To operationalize this principle, ask the following questions:

- Do we have a governance model that helps monitor the impact of the technology on human rights?
- Have we assessed the impact of our new technology on users and their ability to control the application?
- Have we decided who will be in control of the technology and how this control is exercised?
- Have we set a clear accountability process for customers to use to complain or challenge the technology?

The Principle of Explainability

Explainability refers to the ability of the organization to provide a meaningful explanation of the decision made by the AI system. Explainability is a critical concept for establishing trust in AI systems. It provides openness and transparency in the decision-making process. Explainability is also important because it provides the means to those affected by the AI decision-making process to challenge the decision, which helps the system to be more transparent and fair especially when the wrong decision generated by the system is corrected. The ISO/IEC 42001 Standard defines explainability as the ability of the organization to explain to those affected by the decision so people can see the input, the data, and the information that went into the decision-making process. The goal, of course, is to help stakeholders and those affected by the automated decision to understand in a meaningful way, how the decision was reached, and the different factors that went into the decision.

But what does explainability entail? And how is it different from interpretability? And how should we build an explainable system?

A great framework to understand the issues of explainability is the work conducted by authors Phillips, J., Hahn, C., Fontana, P., and Yates. A., Greene, K., Broniatowski, D., Przybocki, M. (2021). According to the authors, explainable AI requires four principles.

- **Explanation**: The act of providing evidence and reasoning about how the system reached a specific decision. This could include what went into the decision, but also what didn't go into the decision.
- **Meaningful**: The process by which users understand the system's explanation. What constitutes a meaningful explanation may be different for different stakeholders. A programmer may understand the decision in different meaningful ways than the buyer of the final product.
- **Accuracy**: For a decision to be explainable, its explanation has to be accurate. Accuracy of the explanation, or the lack thereof, according to the authors, does not affect the explanation and the meaning of the final decision. The accuracy here applies only to the way the explanation was provided, regardless of whether the decision made sense to the user or not.
- **Knowledge limit**: According to Phillips et al. (2021), the fourth principle of explainable AI is the concept of knowledge limit. What this means is that for an output to be explainable, it has to limit itself to the knowledge that was supplied to it. In other words, a decision cannot be explainable if it was built on knowledge that was out of the scope of the decision.

Furthermore, it is important to mention that the ISO/IEC TR 24028: 2020, Overview of Trustworthiness in Artificial Intelligence, published in 2020, introduces many concepts that can be of help to the organization in building a trustworthy AI. The notions of *ex ante* explanation and *ex post* explanation are good to have in mind when dealing with explainability. In the *ex ante* (the before explanation), the goal is to explain what the system is meant to do and how it intends to behave. The goal is to establish trust in the system before it is used. In the *ex post* (the after explanation), the goal is to justify the decision, the outcome, and the result so the user understands the why of the decision.

The ISO/IEC TR 24028: 2020 also introduces other concepts such as the causal explanation (cause and effect explanation), epistemic explanation (how do we know what we know), and the justificatory explanation (the explanation that justifies the results). This Standard is very helpful to read as it provides definitions and explanations on how to build a trustworthy AI system.

Finally, and to make these principles applicable, the High-Level Expert Group on Artificial Intelligence set up by the European Commission provides a list of seven requirements that help organizations and stakeholders build and implement a trustworthy AI. These requirements are:

- **Human agency and oversight**: The need for AI systems, especially the ones that are classified as high risk to be under the control of humans.
- **Robustness and safety**: AI systems must be reliable in their performance and robust regardless of the circumstances under which they perform.
- **Privacy and data governance**: The need for AI systems to respect privacy and address privacy concerns during the data processing phase.
- **Transparency**: The need for the automated system to be open. Sharing information about the model, the data, and the training methods provides information to the external world about how the system works.
- **Diversity, nondiscrimination, and fairness**: AI systems must be inclusive during their design and must address bias so stakeholders can trust them.
- **Societal and environmental well-being**: AI systems must have a positive social impact that allows for AI to be accepted and sustainable over time.
- **Accountability**: Organization, management, senior leadership, and governing bodies must be accountable when the AI system misbehave. Companies must report negative impacts.

The Principle of Robustness

In computer science, the term *robustness* refers to the ability of the computer program to perform in a consistent way over and over, regardless of the changes happening in the environment. It is a central concept of modern technology as it helps build trust in the system. When a machine behaves accurately and consistently in the lab and in the real world despite what might happen in its environment, that machine has met the requirements for the principle of robustness. When you hit the brake and the car stops regardless of the circumstances, your braking system is robust and that helps you build the trust and the confidence that makes you derive and speed with fewer worries.

Robustness is a key concept in AI and intelligence systems in general because of the learning capabilities that intelligence systems may develop over time. So, to build an AI system that is safe, secure, and trustworthy engineers must work on the robustness principle to minimize variability and reduce potential risks and harms that may be associated with unknowns of the real world. This is why a human-in-the-loop approach is a vital approach in building AI systems as it allows engineers to check and identify the variability in performance between the behavior of the model in the lab and its behavior in the real world over time.

TOOLBOX 13

RESPONSIBLE AI: HOW TO DEPLOY A TRUSTWORTHY AI SYSTEM

In the previous section, we explained the process by which an organization analyzes the context. We focused on the legal and ethical dimensions because these are the elements that present a higher risk for the organization in terms of harm, liability, and reputational risk.

However, managers can do more to build a trustworthy and responsible AI. As a general-purpose technology, AI will have a deep impact on society that goes beyond what we have seen so far. It is a technology that is going to be around, and its mass adoption is only a question of time. So, organizations must set the foundations that will allow this technology to gain the social trust it needs to grow.

Many frameworks can be used to build a trustworthy and responsible AI. ISO/IEC TR 24368:2022, Overview of Ethical and Societal Concerns, published in 2022, provides helpful tools and methods that allow organizations to delve into the issue of trustworthy and responsible AI, and how it should be operationalized.

If you want to build a trustworthy system, start by asking your team the following questions:

- Have we identified the group of stakeholders that the AI systems may affect? And have we categorized them?
- Have we established a system by which we classify AI systems by degree of risk and level of impact on users' health, safety, and well-being?
- Have we identified the vulnerability of AI systems, and how these vulnerabilities are dealt with?
- Is our AI ecosystem transparent to stakeholders, and can we open our algorithms, data models, features, and training processes to external parties?

ISO Certification Process: How to Conduct an Analysis of the Context

The standard requires that you analyze the context of the organization, but it does not prescribe a specific methodology. While most organizations do conduct this type of analysis frequently, for ISO/IEC 42001, it is recommended that you conduct a new analysis of the context of your organization. Given the fluidity of the regulatory environment in which AI is evolving and the speed at which the technology is moving, it is recommended that you start from scratch. Avoid simply updating or refreshing old and existing documentation of organizational context analysis.

There are many methodologies available for organizational analysis. You can use any of the known methodologies, such as SWAT analysis (Strengths, Weaknesses, Opportunities, and Threats), the PESTEL methodology (Political, Economic, Social, Technology, Environment, Legal), Scenario Planning, Stakeholders Analysis, Porter's Five Forces, or simply the Industry Analysis methodology. As long as you are asking the right questions, documenting the insights, aligning those insights with your AI strategy, and showing auditors how your organization meets the standard's controls, your

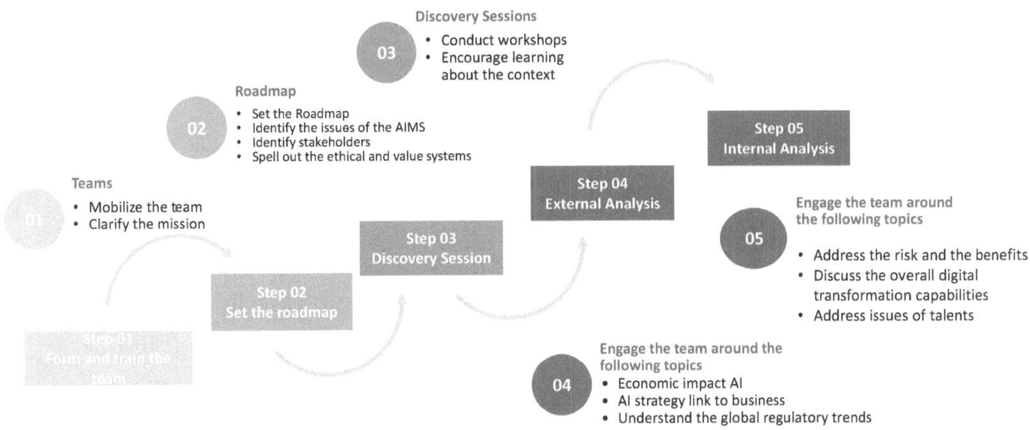

Figure 5.4 ISO certification process: how to conduct an organizational context analysis.

analysis will support the certification process. Remember, the goal is to learn from the analysis about the trends impacting the development of AI technology, and how those trends affect the building of your AIMS, so you can document and justify your choices to auditors. You can also use the organizational analysis process for benchmarking purposes to learn about best practices that help you design a long-term sustainable and viable AI management system in an environment that is characterized by policy uncertainty and legal ambiguity. Figure 5.4 provides a summary of the organizational analysis process according to ISO/IEC 42001.

Following are the steps that you need to take.

Step 1: Mobilize the Team and Clarify the Mission

The first step is to put together a team. Choose members who are well aware of the organizational strategy and have a deep knowledge of the industry, its directions, and the global trends affecting technology. In putting together the team, it is critical that you diversify your team to include many functions, expertise, and backgrounds. Go beyond the obvious functions, such as IT and strategy, to include legal affairs, compliance, HR, customer service, marketing, sales, purchasing, quality, and teams that work on technology or interact with technology vendors.

Step 2: Set the Roadmap

Then you need to clarify the purpose and mission of the team and make sure that the team members are aware and have read the requirements listed in Section 4, Context of the Organization of the Standard. ISO/IEC 42001 lists four reasons for which an organization analyzes its context. These reasons are:

- Identify the issues the organization faces concerning building and developing an AIMS, so the organization can achieve its strategy.
- Spell out the role of the organization concerning AIMS. ISO/IEC 42001 lists all possible roles that may apply to organizations in paragraph 2. But you can also check another ISO Standard, ISO/IEC 22989:2022, Artificial Intelligence Concepts and Terminology, which provides a broader overview of different roles, their definitions, and how they should apply to your AIMS. You may find your organization assuming more than one role. You may think of your company as a data producer but your company is also an AI provider.
- Spell out how your AIMS affects and interacts with your stakeholders, locally and globally. Another ISO standard you should look at during this process is the Standard ISO/IEC 29100:2011, Security Techniques and Privacy Framework. This standard provides a specific framework and terminologies that relate to privacy and security.
- Identify the national and international legal framework, policies, and overall values and norms that may affect your AIMS. Many national legislations have been already enacted, as I explained in my previous section.

Step 3: Conduct Discovery Sessions

Once you put the team together, and you understand the purpose of the context analysis, it is important that you organize workshops and discovery sessions to discuss the blueprint of the analysis. The goal is to have all team members in charge of the AIMS contribute to the conversation. This is not only an analysis of the context for an ISO audit, but rather a learning opportunity for managers and employees to get acquainted and updated on a technology that is moving fast.

Step 4: Start with the External Environment

Ask the following questions to generate insights and make sure you record those insights in a document. This is the document that you will show to the auditors.

- What economic impact AI is expected to have on our industry, and how should we position ourselves to take advantage of it?
- How does AI strategy fit into our overall business strategy, and how do we create coherence and alignment between the two beyond the automation imperative?
- What are the external factors that will enable us to achieve our organizational goals?
- What are the regulatory environment and government's policies that we need to address?
- How does AI development affect regulation about employment, discrimination, customer protection, health and safety, and other matters?
- How has AI impacted public opinion and what are the new emerging customer's attitudes that may affect our market position or AI products?
- If we are using third-party AI products, have we audited those products to make sure they comply with laws and regulations and subscribe to our ethical AI framework, and the generally agreed-upon principles recognized by ISO standards and other international organizations such as OECD and UNESCO?

Step 5: Conduct an Internal Analysis

Once you finish the analysis of the external environment you turn to the analysis of the internal context. In the internal context, you can also take advantage of some of the known analytical frameworks, such as the Balanced Scorecard, the Value Chain Analysis, and the Resource View methodology. The goal is to ask questions that generate more insights about how AI impacts the overall strategy and how the organization intends to create a coherent AI management system that is aligned with the overall strategy and in coherence with other management systems.

Here are some of the questions that you can use to capture insights from the team.

- To what extent AI will affect our digital transformation strategy, and what are the risks of this technology? Risks that relate to the technology itself, but also to the speed at which we are adopting this technology?
- How has the introduction of AI tools and AI products, including third-party products, affected our mission, vision, and values?

- How has AI affected our talent management strategy and how to future-proof our talent strategy is?
- What are the employees' upscaling initiatives that we need to launch to adjust our human capital to this technology?
- To what extent do we need to rethink and reengineer our processes to accommodate the standard requirements?

Chapter 6

Clause 5: Leadership

PURPOSE

The purpose of this chapter is to highlight the role of leadership in building, maintaining, and sustaining the AIMS and explain the certification process.

OVERVIEW

The main focus of this chapter is to discuss the role leadership plays in building the AI management system. As a new management system, AI management system requires resources and engagement, guidance, and a vision that only leaders can deliver. A clear commitment of leadership is *sine qua non* condition of the success of the AI management system. You will learn how to develop an AI policy, what elements should be included in the AI policy, how to assess the competitive landscape to come up with an AI strategy that supports your business strategy, and, finally, how to reorganize internally to develop AI use cases for your AI strategy. The goal is to help you put together the documentation needed for leadership engagement to meet the ISO certification requirements.

DOI: 10.4324/9781003463979-8

BENEFITS

At the end of this chapter, you will be able to:

- Understand the role leadership plays in setting up the AIMS.
- Develop a roadmap for your certification process.
- Understand what the ISO standard requires of leaders to put together the AIMS.
- Develop an AI policy that meets ISO requirements.
- Identify the AI use case that fits better your AI strategy context.

Introduction

The engagement of leadership in building the AI management system (AIMS) is critical. An explicit commitment from leadership that defines the North Star, sets the tone, and creates the incentives that engage people is of paramount importance. In an environment characterized by fast-moving technology and regulatory uncertainty, the role of leadership in guiding the teams in charge of deploying and implementing AIMSs is vital.

According to the ISO Standard 42001:2023, there are three main actions leaders need to take to build a viable AI management system.

- Set the vision and the priorities and show engagement in building the AI management system.
- Develop an AI policy that defines the AI intent, the rules and procedures governing the AI management system, and other processes. The policy should be an auditable document.
- Define roles, responsibilities, and the authorities of different actors engaged in the making of the AI management system.

In the following sections, I will go over these requirements. I will explain them, show you how you put them into practice, and share some best practices from leading organizations.

Set the Vision

AI is a transformative technology. I argued earlier that the impact of AI on the economy and the wider society is just being felt. The emergence of generative AI and Large Language Models has ushered in a new era of disruptions that will be felt by all sectors of the economy. As said by Bill Gates, the age of AI has begun, and it is creating a new reality that we have not seen before. More importantly, the environment surrounding AI evolution is constantly shifting. The technology itself is moving fast, the social interest in the technology is growing, and the calls to regulate this new technology are intensifying. AI has suddenly become a topic of national conversation and calls to limit its impact have grown.

In this shifting environment, a vision of what AI means for the organization is vital. The vision guides employees and sets goals for the organization so the entire organization is aligned with the same goal. A vision inspires people and rallies them behind the AIMS while sharing with society and the general public the dream that the organization wants to achieve. Especially true during uncertainty, a vision provides a line of sight on where the organization wants to be.

But how do you set an AI vision? And how do you create an engaged leadership? What are the steps that you need to take to build the leadership capability required for such transformative technology?

The ISO Standard 42001:2023 lists a set of actions that leadership should engage in. Below I will discuss the most critical ones that have a direct impact on the viability of the AI management system.

TOOLBOX 14

DOES MY COMPANY HAVE AN AI VISION? ASK THE FOLLOWING FIVE QUESTIONS

To help your company develop an AI vision, and to test your leadership on whether they know the vision or not ask the following five questions:

1. Have we clearly articulated our AI vision, and have we defined our aspiration for what AI will do for us?
2. What do we want to achieve in the AI strategy and how should we challenge the status quo?
3. To what extent the AI vision has been communicated to all stakeholders internally and externally?
4. To what extent and in what way employees can align their behavior with the AI vision?
5. In what way does the AI vision help structure organizational behavior?

Set the Vision, Define the Priorities, and the Strategic Direction

The role of leadership in defining the priorities is important as it provides the organization with clarity on where the organization wants to go. This AI vision should be in coherence with the organization's vision, mission, and values, and should send a clear message to all management levels about the importance of the AI management system and how this system helps the organization achieve its business goals.

Defining the vision and how this vision breaks down into strategic goals and strategic KPIs helps you design a viable AI strategy. Organizations may approach AI strategy in different ways, depending on the economic sector and the industry in which they operate. But generally speaking, companies integrate AI to automate processes and decision-making and use it as a platform to link key functions.

A critical element to the setting of an AI vision and strategy is to embrace a holistic view of AI strategy, by integrating the AI into the digital transformation strategy. Decoupling the AI strategy from the overall digital transformation strategy the company is going through weakens the AI strategy and creates more dissonance in the AI management system. A truly successful AI strategy is going to be linked and aligned with the operating model, the data infrastructure, the business model, and the overall organizational plan. Furthermore, leadership needs to engage to make sure that AI strategy and the building of the overall AI management system should not be approached as a purely IT issue. The nature of a general-purpose technology makes AI an enabler that cuts across many functions, teams, and capabilities. Teams need to be aware of this imperative, and it is the leadership role that sets this tone.

TOOLBOX 15

SAMPLE OF AI VISION FROM LEADING COMPANIES

Conduct a search on the top 10 technology companies to see how they crafted their AI vision and mission. Read the vision and mission statement. Notice that the vision should be clear, concise, to the point, and carry a powerful message that engages employees and makes customers believe or curious about their product:

1. Microsoft: "To empower every person and every organization on the planet to achieve more through artificial intelligence."
2. IBM: "To create a future where every decision, every business process, and every experience will be augmented with artificial intelligence."
3. Google: "To organize the world's information and make it universally accessible and useful with the help of artificial intelligence."
4. Amazon: "To use artificial intelligence to make every customer interaction more natural and intuitive."
5. Facebook: "To build technologies that help people connect with friends and family, find communities, and grow businesses using artificial intelligence."
6. Tesla: "To create a world where cars can drive themselves safely and efficiently with the help of artificial intelligence."
7. NVIDIA: "To advance artificial intelligence and create intelligent machines that can perceive, reason, and learn."
8. Salesforce: "To deliver the world's most intelligent customer relationship management using artificial intelligence."
9. Intel: "To develop artificial intelligence technologies that can transform industries, solve global challenges, and improve people's lives."
10. Siemens: "To use artificial intelligence to transform industries and help solve some of the world's biggest challenges."

Lead with Responsible AI Principles (RAI)

The transformational nature of AI requires a transformational leadership style. Leaders should engage the entire organization through a wider transformation strategy that helps employees understand the need to change, the scope of change, and how the results of the change will impact them. In the change plan, leadership should communicate key factors required for the successful implementation of an AI management system. The emphasis should be on how responsible AI principles should help the organization achieve its business goal, act responsibly, and achieve its compliance and certification requirements.

Responsible AI is a framework that helps you design and implement AIMS by the organization's ethical framework and the emerging AI ethical principles we discussed in the previous chapter. In simple words, a responsible AI approach means that your AI strategy and the overall AI management system should be designed with good intentions, and in compliance with laws and regulations while subscribing to ethical AI principles, such as do no harm, fairness and discrimination, human oversight and respect for human autonomy, and explainability.

The Standard ISO/IEC 42001 is the first requirement standard that puts the RAI concept into practice. Adopting RAI principles has many benefits to the organization as there is a real business case to be made for the organization that adopts RAI principles. Many reasons explain that:

- Organizations that create a better trustworthy relationship with the customer and stakeholders benefit their shareholders and employees.
- Organizations that use RAI throughout the product development cycle gain better product quality which in turn improves their customer retention and engagement.
- Organizations that develop a data management process tend to explore better-quality data that is reliable and that itself leads to better decisions.
- Organizations that use RAI principles have better chances of attracting talent. The value system and the ethics of companies are becoming a deciding factor for many hard-to-find talents.
- As regulation is looming and the call for an aggressive regulatory framework intensifies, organizations that adopt the RAI principles will have a first-mover advantage. These organizations will be ahead of the curve in terms of compliance and risk management.

The Responsible AI Institute (RAI Institute), a nonprofit organization, and a member of the World Economic Forum Global AI Action Alliance, is among the first industry coalitions that have developed a comprehensive certification program to address responsible AI. The RAI Institute framework is made of six building blocks: Systems and operations, explainability, interpretability, accountability, consumer protection, bias and fairness, and robustness. This framework not only builds on the many laws and regulations published but also takes into consideration best-emerging practices suggested by organizations such as the ISO, IEEE, UNESCO, OECD, and standard bodies, such as ANSI, BSI, and SCC, as well as many organizations in the private sectors such as IBM, Deloitte, EY, Microsoft, BCG, PwC, and McKenzie.

TOOLBOX 16

In July 2023, the White House announced that it secured a voluntary commitment from leading AI companies to develop a framework that would help them manage AI risks and accelerate the path toward Responsible Artificial Intelligence. These companies, which voluntarily committed to the White House proposal, are Amazon, Anthropic, Google, Inflection, Meta, Microsoft, and OpenAI. This commitment is built around three pillars: Safety, security, and trust. Following is a summary of the main points of this commitment:

- **Sandboxing**: The companies commit to testing their systems internally before releasing them to the public. The testing should be carried out by third-party independent experts.
- **Best practices**: The companies commit to sharing information and knowledge across the industry and with different stakeholders, such as academia and the government, about AI best risk management practices.
- **Cybersecurity**: Companies commit to invest in cybersecurity to protect models and proprietary data that serve to build these models.
- **Incident reporting**: The companies commit to working with third-party experts to discover and report vulnerabilities in AI systems after they are released to the public.
- **Watermarking system**: The companies commit to developing mechanisms that help users know when they are interacting with AI-generated content.
- **Prioritize societal risk**: The companies commit to prioritize research about the most critical harm AI systems pose to society, including bias, discrimination, and protection of privacy.

For more information, visit this link: https://www.whitehouse.gov

Set the Tone and Use Proactive Communication

When embarking on a new technology, communication is critical. Especially true for a nascent technology that is also controversial, the setting of the tone is the essence of leadership commitment. The leadership role in framing the issue of why the organization needs to invest in an AI management system, how it should build a long-term vision for the AIMS, how it intends to engage with different stakeholders, how it navigates the known and unknown risks surrounding the development of AI technology, and how it applies RAI principles creates a change momentum that rallies employees around AI. Leaders should explain the risks associated with nascent technology investments, emphasize responsible AI principles, and choose the right language and vocabulary to set the tone. They need to make compliance with ethical AI a priority and demonstrate how they intend to operationalize it. Always explain the context of the technology so people can see where you stand on difficult decisions. Tell the story of the moral and ethical dilemma leaders go through when deciding about risky technology investment. Finally, be humble and acknowledge the uncertainties of the technology so you can build trust and get the engagement you need from your employees.

AI Policy: Characteristics and Components

In addition to the AI vision that should provide the organization with clarity on where the organization stands concerning AI, Standard 42001:2023 also requires that you craft an AI policy. This is an official company document that captures the commitment and engagement of leadership to the AI management system. If the AI vision helps organizations and employees understand the purpose of AI, the AI policy serves as a reference document that defines procedures and guide behavior, so employees stay compliant.

According to the ISO/IEC 42001, the goal of an AI policy is to define the AI management system, identify the roles and responsibilities of the different people engaged with the AIMS, and show how the AI management system integrates into and interacts with other management systems.

Below I will show you how you develop an AI policy. This will be a generic template that you can adjust to your industry. Then, I will suggest a process by which you can create this policy.

What Should Be in the AI Policy?

Depending on the sector and the industry in which you operate, and the legal jurisdictions in which you are located, your AI policy should include the following elements.

A Statement on the Scope of the Policy, Its Purpose, and What the Policy Intends to Achieve

This is generally accomplished by a small paragraph that is based on the strategy of the organization and other policies, like the data policy, the privacy policy, the HR policy, finance and accounting policy, purchasing policies, and other policy documents that may exist in the organization. Explain the scope of the AI policy, the objectives, and the requirements.

Guidelines on the Use of AI in the Organization

This is usually achieved by defining what is an acceptable use of AI applications, and what is not. Here you define the boundaries of the AI applications. Refer to this paragraph for laws and regulations, standard requirements, and compliance. Explicitly define what is prohibited and what is not and address the gray areas.

Show How AI Management System Integrates with Other Management Systems

This is a crucial element of the AI policy. This paragraph helps employees identify connections, interactions, and the links the AIMS has with other management systems. For most organizations, this would be a challenge. Adopting a holistic perspective, for a nascent technology that is cutting across all business functions, is a big hurdle. The risk of not identifying these connections will create misalignment and dissonance in the overall management system.

Define the Roles and Responsibilities

In this section, you need to identify the roles and responsibilities of different actors engaged with the AIMS. Keep in mind that while ISO/IEC 42001 does not define these roles there is another ISO standard, ISO/IEC 22989:2022 Artificial Intelligence concepts and terminology, that provides definitions of different roles. Depending on the context of your organization, I suggest that you define roles and responsibilities by type of role: Technical roles and business roles.

Technical Roles

- **Data scientist**: Engineer who processes data. This is a critical role to define because data scientists have a deep impact on data collecting, data sampling, and model training. Typically, they are not domain experts, so they should be trained in domain-specific knowledge.
- **Machine learning engineers**: Engineers in charge of AI model implementation, including the production and testing of the model's validity.
- **Domain expert**: Experts who have specific knowledge about a particular field. These experts need to work hand in hand with data scientists and ML engineering to guide them into the particularity of each field such as marketing, customer service, HR, and procurement.
- **AI designer**: Employees who serve as interfaces between data scientists, ML engineers, and users. These are the translators of AI systems into a user-friendly application that impacts the final evaluation of the product.

Business Roles

- **Executive leadership**: Senior leadership in charge of the overall strategy and the direction of the organization.
- **Project managers**: Managers in charge of leading and implementing different AI projects.
- **Compliance managers**: Experts with legal and standardization backgrounds that guide the organization on compliance issues.
- **AI strategist**: People with technical and strategy backgrounds. They have a good understanding of AI trends and how these trends affect the industry and the company.
- **Ethics committee**: This is a new and important role that is directly linked to the development of AI. This committee oversees defining policies, developing frameworks, reviewing risky applications, and providing recommendations on how to manage and mitigate risks.

Data and Privacy

Typically, organizations should already have an existing data and privacy policy that you can link to the AI policy. But if the organization does not have these policies, create new ones. In these documents you typically define the process by which you manage, store, and secure data, and how you handle, share, and protect personal information.

AI Compliance

This is a paragraph that defines compliance and how the organization conforms to laws, regulations, and standards. It may be good to list the set of ISO and industry standards that you comply with and outline ways of addressing noncompliance issues.

AI Talent Management

A good policy should include a paragraph on how the organization intends to train and develop employees in charge of AI systems. This should also link to the HR and training management system.

Monitoring and Improvement

An AI policy should also include a paragraph on how the organization addresses the improvement of the AI management system. This is a control of ISO/IEC 42001.

Review and Alignments

Finally, the policy should align with other company policies and should be reviewed to ensure it takes into consideration the change in the technology and the regulatory environment. The review and the alignment with other policies are a requirement and part of the controls listed in Annex A. Figure 6.1 provides a summary of the AI policy components.

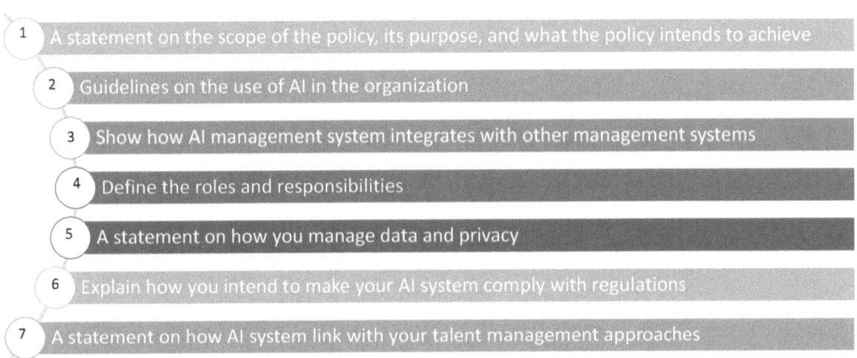

Figure 6.1 Components of an AI policy.

How Do You Create an AI Policy?

Following are the steps that you need to take to create an AI policy.

Form the Team

As we argued earlier, teams in AI should be as diversified as possible. Put a cross-functional and multidisciplinary team that has a range of expertise, such as policymaking, strategy, data science, marketing, compliance, legal, and HR. Depending on the sector in which you operate, your policymaking team should reflect those who will directly or indirectly interact with the AI management system.

Engage with Stakeholders

To ensure that the policy is as comprehensive as possible, you need to engage with your stakeholders such as employees, customers, partners, and vendors. The goal is to hear their perspective on how AI will impact them.

Conduct Discovery Sessions and Workshop Meetings with Different Stakeholders

To ensure that you are capturing the views and the perceptions of different segments of the stakeholders, conduct discovery sessions. Ensure that the discussion is open and covers a wide range of topics.

Review the Laws, Regulations, and Ethical Framework

Your legal team should draft a manual of all the laws and regulations that apply to AI. This will guide your discussion and should help members of the team understand the risks and liabilities of AIMS. Take a look at the following ISO standards as they address, directly or indirectly, the issue of governance in AI, and review the legislation I outlined in the previous section.

- ISO 37000:2021. Governance of organizations – Guidance
- ISO/IEC 38500:2015. Information technology – Governance of IT for the organization

- ISO/IEC 38505-1:2017. Information technology – Governance of IT – Governance of data – Part 1: Application of ISO/IEC 38500 to the governance of data
- ISO 31000:2018. Risk management – Guidelines
- ISO/IEC 23894:2023. Information technology – Artificial intelligence – Guidance on risk management.

TOOLBOX 17

HOW TO GAUGE YOUR LEADERSHIP ENGAGEMENT? ASK THE FOLLOWING FIVE QUESTIONS

How do I know that leadership is truly engaged in supporting and promoting AI management system? If you want to check whether your top management is truly engaged in supporting and developing the AIMS, ask the following five questions:

1. What actions or steps our top leadership has been involved with to define and develop a common language and framework around AI policy and AI strategy?
2. Have we defined the level of leadership that owns AI strategy?
3. To what extent our top leadership is engaged in leading our AI ethics?
4. What actions or steps our top leadership has taken to drive engagement around AI policy?
5. How and to what extent our top leadership is engaged in creating and fostering a risk management-based culture?

AI Strategy

An essential step in the creation of a successful AIMS is the definition of a clear strategy on how the organization intends to use AI. While AI strategies may look different from one organization to another, depending on the industry and the economic sector where the organization operates, the process of formulating a strategy is relatively similar. For an AI strategy to be effective, I suggest the following three steps:

Step 1: Develop AI Use Case

Enhancing Customer Satisfaction

Getting closer to your customers and knowing about their needs and wants is critical to building a long-term relationship with your customer base. The predictive power of intelligence systems, such as conversational AI, presents a strong use case. The power of AI and its ability to identify patterns find connections between customer's clicks and behaviors, and make recommendations is impressive. Response time to customer inquiries is also improved when you introduce AI. A survey by Sales Force Research found that 51 percent of companies choosing AI do so to personalize the customer experience. Seventy-eight percent of customers, in a survey conducted by Oracle, are more likely to do business with a company that personalizes their experience.

Agile and Data-Driven Decision-Making Process

AI applications allow decisions to be driven by what data suggests. AI provides better and faster decisions. It helps you reduce the time to make complex decisions, which adds more agility and nimbleness to your decision-making process. In a data-driven environment, response time to market changes is faster and more accurate.

Creating Efficiencies

The implementation of AI and the automation of different processes provide a real cost-saving opportunity. Automation relieves employees from mundane tasks, reduces errors, and allows talent to focus on more business-critical issues.

Improving Productivity

Faster processing with analytics and improvement in the accuracy of the data helps the organization improve its productivity by speeding up processes and saving waste time. A study by Accenture found that AI can increase productivity by 40 percent and reduce staffing costs by 70 percent. In the marketing function, a study by IDC found that companies using AI increase the efficiency of their marketing processes by 30 percent.

Step 2: Assess the Competitive Landscape

The second step in the design of an AI strategy is to analyze your competitive landscape. While competitive analysis is a complex task, as it requires information that may not be available to you (information such as competitor's product, quality issues, product positioning, and pricing), it is a critical step in building your AI strategy. The good news about building a strategy for nascent technology is that most of the competition is at the same level as yourself, outside of the course of digital natives. So, conducting this analysis may be easier as you and your competitor are at about the same level.

A framework suggested by the World Economic Forum (2022) provides a good starting point to how you should analyze the competitive landscape. The framework, made of four quadrants, asks two essential questions:

- Will the strategic impact of AI improve operational efficiencies or create a new business model? Here, the issue is to distinguish between a technology that merely improves efficiencies, from a technology that radically transforms the business model.
- The second question of the WEF model asks about the competition. Who is deploying AI? Is it a new player from an adjacent industry or a competitor?

The World Economic Forum framework is a great tool to help you analyze the competition, assess the risk, and decide on the design of the strategy. An extreme scenario, in the WEF model, is a scenario in which you are facing a newly entrant competitor introducing a completely new business model (a scenario that will be in the upper-right quadrant of the WEF model) that represents more risks and requires swift actions.

Step 3: Reorganize Internally

Designing an AI strategy requires the building of many capabilities that may or may not exist in your organization. A common mistake in formulating an AI strategy is to be tempted to fast-track the AI strategy and thus build it on top of existing capabilities that may not be suited for a powerful and transformational technology such as AI. Three critical capabilities need to be built or radically improved to create viable AI technology.

Build and Update the Current Technology Infrastructure to Empower the AI Management System

This may include new software, new cloud technology, and other tools and platforms that accommodate AIMS. You cannot build powerful technology such as AI on an existing old infrastructure. Leading organizations tend to invest in new infrastructure with end-to-end platforms that link different business functions and applications.

Design a Data Strategy

A good AI strategy requires a good data strategy. There is no AI strategy without a data strategy that defines the process by which you manage, store, and process data. The backbone of an AI strategy is a data strategy that is comprehensive, exploitable, and provides a high level of confidence in its accuracy.

Talent Strategy

According to McKenzie's 2022 AI Global Survey, leading companies in AI strategy are the ones that have invested in building a comprehensive talent strategy. Without a vision of how to upskill, reskill, and improve the core of AI talent competencies, such as data scientists and data engineers, AI strategy will stumble. More importantly, a change in the way teams communicate and work is also a critical component. Because AI cuts across many functions and teams, the silo culture needs to be replaced by a culture of communication and collaboration that breaks down barriers between people working on similar and different issues.

TOOLBOX 18

FIVE QUESTIONS TO ASK ABOUT AI STRATEGY

1. Have we defined our AI strategy and how it fits into our overall corporate strategy?
2. How often and to what extent does our AI strategy get revisited and updated to reflect the external environment, market evolution, and new regulations?

3. Have we spelled out structures and governance processes to support our AI strategy's activities?
4. Have we defined who owns different AI activities and programs and how these programs and activities support the overall AI corporate strategy?
5. Have we established metrics to measure our AI activities and programs?

AI Oversight: The Role of Board of Directors

As a governing body, the Board of Directors has a direct responsibility for how AI systems are used and the risks they produce for the organization. In addition to the ISO/IEC 42001, which addresses management and leadership issues, there is another ISO Standard, ISO/IEC 38507:2022 Governance Implications of the Use of Artificial Intelligence by Organizations, that directly addresses the issues of governance and oversight by Board of Directors. Clause 6.2 (Governance oversight of AI) of this Standard puts the liability of misuse of AI systems on the shoulders of different governing bodies, including the Board of Directors.

A look at how the Board of Directors deals with AI reveals that most of the Board of Directives have not yet set the process by which they define their engagement with the oversight of the AIMS. A survey conducted in 2022, by the Institute of Directors (IOD) a UK-based organization for Directors, shows that 80 percent of members surveyed did not have a system in place to audit AI. More importantly, the survey reveals that the reason for this lack of oversight is that board members did not know the type of questions to ask to exercise their oversight of AIMS.

This is a revealing case. It shows that most of the Board of Directors are not ready yet for AIMS oversight. It also speaks volumes about the nature of AI systems. An emerging and powerful technology that is moving faster than our governance system can adjust. Governing bodies, especially the ones that are not involved in the day-to-day operations, tend to be less engaged in general, and in the case of AI, they seem to be completely removed from the process. Lack of technology knowledge, the shifting of the regulatory framework, and the absence of consensus around best practices and frameworks are all reasons that explain this situation.

TOOLBOX 19

HOW SHOULD THE BOARD OF DIRECTORS GET INVOLVED IN AI OVERSIGHT?

The Institute of Directors (IOD) provides twelve principles that should guide the involvement of the Board of Director members in AI systems oversight. According to IOD Board members should do the following:

- Engage in and with the regulatory framework, as well as monitor how laws and regulations evolve in your country and other parts of the world. Board members should also establish an ethical framework for the use of AI that is auditable and measurable. This framework should be ideally integrated into the ISO auditable standards, such as ISO/IEC 42001 and 9001:2015.
- Conduct an impact assessment on how AI systems will affect the stakeholders, including the impact of AI applications on employees and their jobs.
- Align AI strategy with the business goals of the organization. Work to harmonize AI objectives with the vision, mission, and value of the organization.
- Establish an Ethics Committee that is diverse and with veto power to decide about AI applications and their use.
- Document and detect bias in data sources. Remove bias when bias is identified.
- Build employee AI capacity and empower employees to learn how to use AI and how to report harmful outcomes of AI systems.
- Use a privacy-by-design approach to protect the privacy and the data of AI systems. Train data scientists and engineers producing ML models on ethical AI principles.
- Provide a secure environment to enhance the safety and security of AIMSs to identify vulnerabilities and prevent cybersecurity attacks.
- Before full implementation, the Ethics Committee should provide recommendations on how the system impacts stakeholders.
- The ethics committee should approve and must have the final say on AI implementation. The Board of Directors is directly liable for AI applications.

- Review and monitor the AI systems regulations to ensure that the AI systems are working as it was intended.

This set of principles provides a strong framework for the Board of Directors to engage with AI systems. They also comply with the spirit of ISO/IEC 42001 and subscribe to the responsible AI principles we discussed earlier. It is a good first step for a critical governing body, such as the Board of Directors, which tends to face criticism when things go wrong.

Chapter 7

Clause 6: Planning

PURPOSE

The purpose of this chapter is to explain the planning process by which you create an AI risk management framework. The chapter outlines the steps and the actions you need to take to comply with the certification process.

OVERVIEW

The purpose of this chapter is to explain the planning process required by the ISO/IEC 42001 Standard. The chapter outlines the steps and actions you need to take to comply with the certification process during the planning process.

A central concept to this chapter is the concept of risk analysis, risk treatment, and impact analysis. The chapter explains the planning process that helps you put together a risk management framework and what steps you need to take to classify AI risks. The chapter discusses a typology of risks that helps develop a comprehensive approach to understanding what risk is, how to treat it, and how to mitigate it. A discussion of data-related risks is also provided to help you develop a better approach to improving data quality. Finally, a discussion of leading change management frameworks is also presented to help you plan better the creation of an AI risk management framework that supports your AI management system certification and accreditation.

DOI: 10.4324/9781003463979-9

BENEFITS

At the end of this chapter, you will be able to:

- Develop a better understanding of the AIMS planning process.
- Understand how the concept of risk plays into the certification process.
- Learn how to identify and classify risks that apply to the context of your organization.
- Develop a risk framework to manage data.
- Learn how to plan for the overall change that your AIMS and your risk framework may produce.

Introduction

The planning section of the ISO/IEC 42001 Standard is the most critical section of the entire standard. With Annex A of the standard, the planning section is the core of the standard as it contains critical requirement clauses that relate to risk and need to be met for an AIMS to be compliant with ISO requirements. This section deals with the risk management framework and addresses three key actions:

- The risk assessment process and how to conduct it.
- The risk treatment options and how to address risk when it is identified.
- The process of conducting an impact assessment of the AIMS.

In the following I will address a few conceptual issues. I will discuss the concepts of risk analysis, risk treatment, and impact assessment. Then, I will provide a typology of risk to help you develop a broader approach to the concept of risk required by the ISO/IEC Standard 42001. I will then discuss the planning process by which you put an AI risk framework and how you manage this framework to comply with the standard's requirements. Finally, I will address a few issues that relate to the data management risk planning process, and how you plan the overall change management approach to implement your AIMS risk management framework.

AI Risk Management, Risk Treatment, and Impact Assessment

To understand Clause 6, we need to go over some of the key terminologies Clause 6 adopts, namely, the concept of AI risk assessment, AI risk treatment, and AI impact assessment.

The Concept of Risk

A key terminology that Section 6 uses is the terminology of risk. While the term "risk" in language tends to convey a negative connotation, the way it is used in ISO/IEC 42001 and other standards that address risk management is different. Risk in ISO documents is used in a broader way to refer to any deviation from the initial objective, regardless of the consequences of this deviation, positive or negative. If the deviation is negative, then it needs treatment according to the risk management process outlined in the standard. If the deviation is positive, then this would be considered as an opportunity that the organization might want to take advantage of.

The Concept of Risk Assessment (Clause 6.1.2)

The process by which an organization evaluates the likelihood and the impact of something going wrong that may affect the performance of AIMS.

The Concept of Risk Treatment (Clause 6.1.3)

Risk treatment simply refers to the actions the organization takes to address the risk. Treatment actions to address the risks can be chosen from the ISO Standard 42001:2023 or simply chosen by the organization itself.

The Concept of Impact Assessment (Clause 6.1.4)

Impact assessment is the process by which organizations assess the possible impact of the AIMS on the users.

From my perspective, and without getting into a discussion of semantics, the **term risk assessment** is a broader concept, while **impact assessment** focuses on the consequences of harm that may occur because of a particular risk, like assessing the harm of a data breach on privacy. In the impact assessment, we address real effects and a real impact that may be harmful, while in risk assessment, we address probabilities of a scenario that may or may not materialize. ISO/IEC 42001 requires both actions, a risk analysis, with its risk treatment options, and an impact assessment that identifies harm to a particular type of user (individual, groups of individuals, or society).

TOOLBOX 20

CONDUCTING A STAKEHOLDERS' ANALYSIS

In planning the implementation of your AI management system, you need to conduct a stakeholder analysis. This is a requirement of Section 4.2 that we have explained in the previous chapter (the standard uses interested parties). The goal is to engage with the people and the parties that interact with your AIMS early on, so you can assess, identify, and mitigate risks. You can conduct stakeholders analysis using the following process:

- Create a list of stakeholders that you think will have an interest in your AIMS or might be impacted by it. Typically, a list of stakeholders may include customers, employees, vendors, regulators, investors, and members of the wider society. Stakeholder groups may have a different level of interest and may be impacted differently so it is critical that you define and quantify if possible, the level of stakeholders' involvement in your AIMS, their interest, and how AI will impact them, so you can identify the impact.
- Identify the concerns and needs of your stakeholders through surveys, focus groups, direct communication, or meetings. Documenting stakeholders' concerns will be of great help during the risk management phase.
- Communicate the documentation on risk to teams involved in the production of AI systems throughout the AI project lifecycle.

A Typology of Risks

As an intelligent system, AI is different from other technologies. Because it is built on machine learning approaches, AI systems can learn and evolve by themselves. This is a serious risk as the outcome of the new self-learned capabilities may deviate from the initial intention, thus creating new risks for which the organization is not prepared.

So, what are the risks that may emerge as a result of AI systems deployment? And how do they affect the organization? To understand this, we need to look at the typology of AI risks. The World Economic Forum (2022) and the Committee of Sponsoring Organization (2020) provide a nice typology that helps you categorize risks and the consequences that pertain to each risk. Below is a discussion of the most common risks.

Performance Risk

The performance risk relates to the way an AI system is supposed to work. This is the most common and general AI risk. When something goes wrong in the performance of an AI system, it may be the result of a flaw in the machine learning model, or in the assumptions that went into building that ML model. A performance risk may be also the result of a typical statistical error that pertains to data manipulation, such as model overfitting/underfitting, or misclassification of the data points. Common approaches to mitigate performance risk are to ensure that the data is of good quality, that the robustness testing is monitored, and that the engineers involved in handling data are well-trained to mitigate this type of risk.

Security Risk

Security risks related to malicious attacks. An AI system may be subject to attacks or hacking that may affect its performance. A security risk can be the result of many factors, such as data poisoning, adversarial attacks, and other cyberattacks that may impede the way a system is supposed to work. Security risks are serious, and their impact goes beyond the AI systems. Having a cybersecurity strategy, upstream leadership involvement, and continuous awareness campaigns that raise the level of employees' cybersecurity awareness is one of the most common mitigating strategies. In McKinsey's 2022 State of the AI Report, only 51 percent of respondents said that they have designed mitigation procedures for cybersecurity risk.

Enterprise Risk

Enterprise risks are the result of misalignment between the AI systems and different parts of the organizations. This could be the result of an absence of alignments, or the lack of supporting capabilities that should enhance the performance of the AI systems, like the lack of talent strategy, or data strategy. An enterprise-level AI risk is more complex. It is not easily identified because it deals with the white space of the organizational chart, and the issue of coordination between and among dispersed units and functions. While this type of risk can be the result of many factors such as size, culture, burnout, and fast changes, it is also indicative of weak leadership oversight.

Reputational Risk

When any of the previous risks happen, it creates an impact on the reputation and the brand name of the organization. This is perhaps the most complex of all the risks, because, once an issue gets in the public sphere, it becomes outside of the control of the organization. When a backlash against a product of the company happens, repairing it takes time, as it takes time to change people's negative perceptions. PR campaigns and apologies do little to repair the reputational damage, and that is why senior leadership needs to get involved in this particular risk, ahead of time. A good mitigation strategy is to make sure that AI governance mechanisms are in place, and that rules and procedures of the AI life cycle project are followed.

Legal and Regulatory Risk

Violation of laws and regulations by AI systems presents a high risk. Particularly true during regulatory uncertainties when legal requirements shift. A good mitigation approach is for the legal team, and the HR team to engage early on during the production of different AI applications and algorithms. For instance, conducting algorithm audits of HR automated decision systems can help the organization avoid discrimination and bias. Only 26 percent of participants in the 2022 McKinsey State of AI Report said that they have built mitigation strategies for legal and regulatory risk.

AI Scalability Risk

A risk that is inherent to the nature of an automated system relates to AI's ability to scale and amplify the decision-making processes. When decisions are made by humans, the traditional way, their consequences tend to be limited in scope and context. When decisions get automated, their impact gets amplified through speed and scale. This has the potential of impacting a larger number of the population and creating a larger risk that may span different legal jurisdictions.

The Black Box Risk

Machine learning models, and particularly deep learning neural networks, use huge amounts of data to produce outputs, without explaining how the model reached its conclusion. Because of the complexity of the data variables an ML model uses to make decisions, the risk of humans not being able to understand how the system reached its decision is high. The lack of explainability and interpretability presents a legal and ethical risk. When senior leadership cannot explain how a system reaches its conclusion, it creates accountability and a trustworthiness problem that may affect the entire governance model. Figure 7.1 provides a typology of the AI risks.

Figure 7.1 A typology of AI risks.

TOOLBOX 21

THE PLANNING PHASE: HOW TO LINK ISO/IEC 42001 TO OTHER ISO/IEC STANDARDS THAT ADDRESS THE OVERALL GOVERNANCE ISSUE

To better understand how to implement the planning section of the 42001, you need to read it in conjunction with two other ISO/IEC standards: ISO/IEC 38507:2022 Governance Implications of the Use Of Artificial Intelligence by Organizations; and ISO/IEC 23894:2023, AI Guidance on Risk Management. The first standard (ISO 38507 governance) presents an overview of the governance process, frameworks, tools, and the general philosophy of how governing bodies should engage with AI systems. It is a guidance standard, not a requirement, that can be of great help to all types of organizations, regardless of their size and industry. It is also linked to other ISO standards, such as the Governance of Organization (ISO 3700), the Governance of IT (ISO 38500), and the Governance of Data (ISO 38505).

The second ISO standard you need to look at during your planning phase is ISO/IEC 23894 Artificial Intelligence, Guidance on Risk Management. This is a critical guidance standard that adds value to the planning process. It builds on ISO 3100 Risk management and provides frameworks, tools, and processes by which an organization should conduct a risk management analysis of the AIMS.

Furthermore, to better implement Section 6 of 42001, you need to read it in conjunction with Section 4 (the Context of the Organization) of the ISO/IEC standard 42001. A critical dimension is to incorporate the results of the organizational analysis you conducted into the implementation process by answering the following two important questions:

- How did your understanding of the context help you make a better plan for the standard implementation?
- What specific issues has the context analysis identified that you need to pay attention to and address during the planning of the implementation process?

Section 6 is linked to Section 4 (organization context). In Section 4, I showed you how to conduct an analysis of the external and internal context of the organization. Here you need to highlight how your understating of the organizational context helped you plan better, by emphasizing the specificity of your context, the legal and regulatory framework, and how they pertain to the implementation of your AIMS.

AI Risk Management Planning: Principles, Framework, and Process

Putting together a viable framework for AI risk management is a complex task. Especially true, for a transformative technology that is rapidly evolving. A survey conducted by Deloitte, *The State of the AI*, in 2022, found that 50 percent of respondents to the survey consider risk as a barrier to scaling AI projects, while only 30 percent of respondents said that they have aligned their AI risk management process with that of the overall organizational risk management framework. Another more revealing finding about the complexity of the risk management process and how organizations deal with AI risks comes from the 2022 McKenzie *State of AI Report*. In this survey, McKinsey did a comparison of risk mitigation strategies between the years 2019 and 2022 and found that little has changed since 2019 in the way organizations deal with AI risk management and mitigation strategies.

Your approach to risk management should be an anticipatory one and as comprehensive as possible. Fortunately, during the last two years, there has been an explosion in AI publications, research, and frameworks that guide AI risk management. Here is a list of helpful frameworks:

- US National Institute of Standard and Technology (NIST) Artificial Intelligence Risk Management Framework (AI RMF 1.0)
- UK Information Commissioner's Office (ICO) Data Privacy Impact Assessment
- COSO Enterprise Risk Management (ERM Framework)
- Responsible AI Institute Framework
- And many other organizations in the private and technology sector such as Microsoft, BCG, Deloitte, and PWC have also published helpful frameworks and tools to conduct AI risk analysis and AI impact assessment.

AI Risk Framework: A Requirement to Certification

According to the McKinsey 2022 *State of AI Report*, interest in AI, as shown by the level of investment and adoption, has accelerated between 2019 and 2022 as more companies report that AI has become a priority for their strategy. However, when it comes to AI risk management, the survey depicts a different portrait. There has been no progress during the last five years in AI risk management strategy, and, in some cases, the interest even decreased.

According to the McKinsey report, the number of companies who have considered AI risks in security in 2019 was 48 percent, while in 2022 that number climbed only moderately to 51 percent (three percentage points increase). For AI risks in regulatory compliance, the progress during the last three years was even less, 35 percent in 2019 versus 36 percent in 2022 (one percentage point increase). In personal and individual privacy risk, the interest was 30 percent in 2019 versus 28 percent in 2022 (a drop in interest by two percentage points). For organizational reputation, the interest in risk was 29 percent in 2019 versus 22 percent in 2022.

These are alarming figures as they show a clear lack of progress in AI risk management strategies despite the accelerated interest in AI investment and adoption. While one would assume that leadership is aware of the implications of this gap on organizational performance, these figures point also to the complexity of the AI risk function in an environment that is characterized by ambiguity and uncertainty.

Next I will discuss the foundations of the risk management framework and provide a few impact assessment tools that you can incorporate into your ISO/IEC 42001 certification process.

AI Risk Management Foundations

A common mistake in thinking about risk management is to assign a person or a department to the function of risk and ask them to come up with a framework to identify, analyze, categorize, and mitigate AI risk. Because AI risk is everywhere, and AI risk tends to permeate all facets of the organization, it is important that, when you address AIMS risks, you take a comprehensive approach. A very helpful risk management framework is the NIST framework. This framework, published in 2023, provides an excellent approach to risks in AI systems. It is also built around a set of principles that are like the principles of ISO/IEC 42001.

AI Risk Should Be Integrated into the Enterprise Risk Management System

AI risk should be part of the enterprise risk management strategy and should be approached as part of management activity. Infusing an AI risk management philosophy into the entire organization raises the profile of AI risk and helps managers and employees engage with the AI risk. De-clustering AI systems risks from your overall risk management practice may negatively affect teams' communication and their common understating of the risk issues, which in turn may affect their engagement in the AI risk activity.

Here are a few questions to ask during a risk assessment process that will allow you to integrate risk into the overall organization strategy:

- To what extent our AI risk management system is integrated into the overall risk management of the organization?
- How does the AI risk management system link to other risks?
- Do we have a central authority that addresses AI risk and risk in general?
- Is there a risk officer in charge of the entire AI system risk across the enterprise?

Embrace a Wholistic Perspective

The second principle of AI risk management, according to the ISO AI risk management, is to structure the risk management activity and to adopt a holistic approach to its deployment. When AI risk activities are structured, they are more sustainable and lead to consistent and coherent outcomes. The comprehensiveness of AI risk activities helps you align risk tools and practices with the vision of the organization. A survey conducted by the *Economist Impact* in 2023 indicates that 77 percent of executives agree on the need to use a comprehensive approach to risk management, 46 percent report investing in an enterprise-wide risk management approach, and 36 percent report integrating risk management into strategy and decision-making processes.

Here are some of the questions to ask to create a holistic approach and structured methodology for your AI risk management system:

- Is our AI risk management process organized, well structured, and comprehensive?
- Does our AI risk system cover all aspects of the AI Project life cycle?
- Does our AI risk management system consider the impact on all our stakeholders?
- Does our AI risk process account for all details, the factors, and the perspectives that need to be addressed, and how comprehensive our risk approach is?

Customize Your Approach

Because organizational context differs from one to another, customization of the approach and the process of risk analysis to the specific context of the organization is of paramount importance. Applying a standardized version of risk analysis may not yield positive results.

Here are some of the questions to ask:

- Does the AI risk management system address the specificity of our context?
- How granular our analysis of the context should be? (The more detail and more granularity the better customization is.)
- Does our AI risk management system take into consideration the type of demographics we serve and the type of data we use?
- Does the organizational analysis take into consideration the type of challenges specific to our sector?
- In what way does the AI risk management system take into account the value system, our specific culture, and the way we do business?

Be Inclusive of Your Stakeholders' View

AI risk management process must be inclusive of stakeholders' views and positions. When an AI risk analysis is inclusive, engagement from different actors internally and externally is achieved. More importantly, engaging inclusively creates more diversity in the way the system is designed, which in turn helps better identify bias and blind spots that may create hidden risks. All groups that may be affected, directly or indirectly, by the AI systems should be identified and their voice should be reflected in the risk analysis process. Inclusiveness also adds a better transparency layer to the AIMS.

Here are some questions to ask to make your AI risk management system more inclusive:

- Have we engaged with different segments of society that may be impacted by the AI systems?
- Does our AI risk management framework cover a wider range of demographics, and have we engaged with them to gauge their perception and views about our AI systems?
- Does our risk management classify impacted populations by degree of level of impact and risk?

Adopt an Agile Mindset

Because of the shifting economic and regulatory environment of AI, and the changing nature of the technology itself, a good AI risk management philosophy is the one that takes into consideration change. Approaching this system with an agile and nimble mindset, able to adapt and adjust when technology shifts, helps the AI risk management system to be more responsive. Questions to ask are:

- Is our AI risk management agile and able to adapt to shifting environments?
- Can our AI risk management system adapt quickly to emerging risks?
- Can our AI risk management anticipate changes that may affect its validity?
- Can AIMS adjust to the shifting social perception of the people we serve?

Spell Out Your Assumptions

Another implication of the changing nature of AI technology is the assumptions under which we build AI systems. When designing an AI system, always assume that the design is to the best of what is available in terms of information, data, and overall knowledge. Things may change and that may affect the effectiveness of the AI systems. Questions to ask are:

- Do we train our engineers to keep track of emerging trends in the way data and information are used in producing AI systems?
- Are we transparent with the stakeholders about the type of information and assumptions we use to build the AI systems?
- Do we share with our stakeholders the level of our confidence in our data and what we think about its quality?
- Does our risk management system address the issue of transparency of data?

Pay Attention to the Cognitive Bias

The analysis of risk is conducted by humans. Perception, interpretation, differences in views, professional approaches, and cultural backgrounds all interfere in the design and deployment of AI systems. This is particularly true for global companies that operate across cultures and cross-legal jurisdictions.

Questions to ask are:

- Does our AI risk management system address the issue of diversity in teams in charge of AI production?
- Do we train engineers in charge of AI systems on cultural sensitivity issues that surround AI systems, data sampling, and data collection?
- Does our risk management require documentation of these types of training?

Learn and Improve

The AIMS can improve only if we can capitalize on previous implementation lessons that document aspects of AIMS performance. Assessing the system and learning from the assessment helps improve your AI risk management process.

Questions to ask are:

- Does our AI risk management system take into consideration the need to improve the process by which risk is analyzed?
- Do we keep track of errors, incidents, and issues that help us improve the system?

Finally, it is important to mention that risk drivers that drive organizational interests in improving risk management approaches differ from one organization to another. In a survey conducted by the *Economist Impact* in 2023, it was found that risks that drive organizations' efforts to improve risk management are technological risk (43 percent), health risk (39 percent), financial risk (41 percent), the human capital risk (39 percent), and operational risks (38 percent). Figure 7.2 provides the eight principles of AI Risk Management.

The Planning of Data Management Risk: An Imperative to AI Management System

Data has always been a contentious research subject. Scientists, researchers, policymakers, and business leaders use data to understand complex phenomena, solve problems, and make decisions. A 2022 CIO survey conducted by GovLoop, a US-based research company, reveals that 86 percent of organizations have either adopted or are considering data-driven projects, while 93 percent of federal chief data officers say leadership relies on data-driven insights to make decisions. Data has become a critical component of the modern management function and the decision-making process.

With AI, data has acquired a new status and a new function. Data is now used to teach machines how to make decisions and train algorithms on how to create decision models. Data is the engine that powers the working and thinking of machines. However, issues of data quality, such as representativeness, bias, and sampling, are still a challenge for most companies. A survey on the State of AI Readiness conducted by McKinsey in 2023 reports that the majority of participants in the survey continue to have issues with their training data. Respondents to the survey identified the following problems with data: Data noise (68 percent), biased data (50 percent), and domain gap (or the lack of representativeness of the data) (48 percent). A similar study conducted by BCG found that 70 percent of companies struggle with the quality of data from sourcing it to validating it and sharing it. This is why ISO/IEC 42001 makes data quality a cornerstone requirement of AIMS.

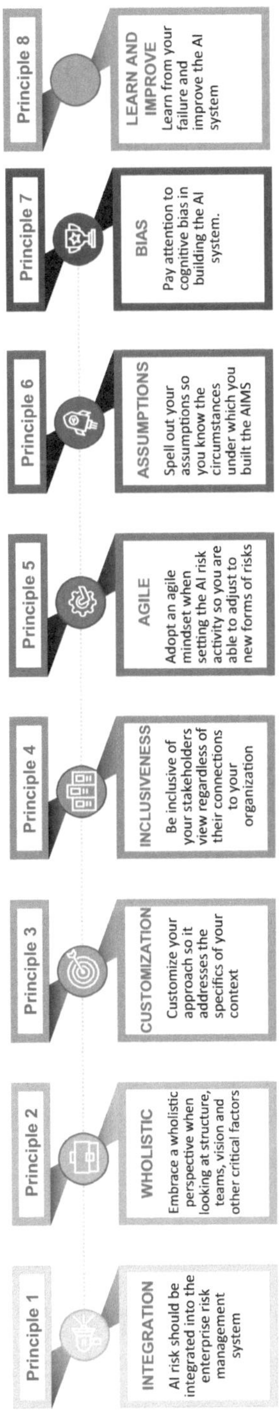

Figure 7.2 The eight principles of AI risk management.

ISO Standard Data Quality Requirements

A crucial requirement for the certification process and the functioning of AIMS is to ensure that the data is of good quality. ISO/IEC 42001 requires that organizations define the requirements of good data quality and ensure that their data management practices meet those requirements (Annex A 7.4).

Good-quality data, according to a Digital Curation Centre report published in 2020, is data that is "fit for use." A fit for use data has the following attributes:

- **Sensitivity**: Data that is anonymized can be exploited commercially and protects the privacy of the individual as well as the causal influences.
- **Representative**: Data that is free from cognitive bias, such as selection bias, exclusion bias, reporting bias, and detection bias.
- **Accuracy data**: Data that is objective, precise, reliable, valid, legitimate, and labeled. In the Zeitgeist survey, 33% of the respondents cite the data labeling as a major issue.
- **Accessibility**: Data that is open, accessible, standardized to a certain degree, and secure.
- **Coverage**: Data that consistently has geographical and temporal coverage. Figure 7.3 provides a summary of the data quality requirements.

But how does data get corrupted? And in what stage of the data collection process does data lose its equality?

According to Karen Hao, an American journalist specializing in AI and contributing to *The Atlantic*, data quality gets impacted during three stages: The data collection stage, the preparation stage, and the stage in which you frame the problem. It is during these three stages that bias seeps into the data. Let's take a look at these three stages.

Data Collection Phase

In the data collection phase, it is important that you pay attention to the way you collect, classify, and label data as bias may seep into the data unintentionally. Bias in this phase can take many forms, like the bias that affects the sampling. If the data you collected is inherently biased toward a group of

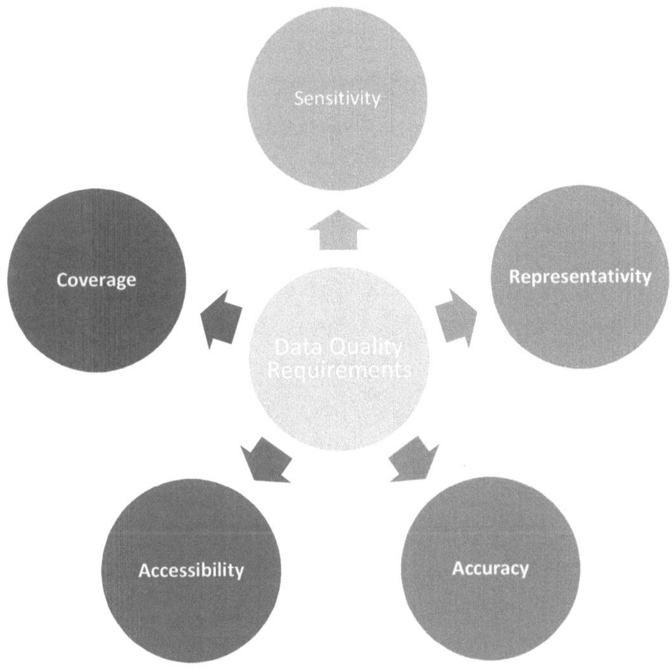

Figure 7.3 Data quality requirements.

people, then your machine learning decision will also carry in them that same bias. In HR, for instance, we use previous hiring practices and promotion data to make predictions about future workforce planning. But if the data you used to train AI to make those predictions about workforce planning is not a true representation of the workforce, then most likely your prediction will be biased against the group that your data did not account for.

Data Preparation Phase

According to Karen Hao, data preparation is an important step in which subconscious bias can be introduced into the AI decision-making process. When we prepare data, we use attributes to segment the sample being studied so we can learn about patterns, correlations, and causation. Examples of attributes may be age, gender, level of education, driving habits, number of traffic accidents, source of income, nationality, zip codes, etc. But if you use an attribute that is in itself discriminatory, then, of course, your algorithm will also learn how to discriminate "subconsciously." A zip code for instance, while on the surface may be a neutral attribute, in reality, it might be a discriminatory factor.

Problem Framing Phase

The third way that might create bias in your AI decision-making, according to Karen Hao, is the way you frame the problem your AI system wants to solve. This is another way through which bias seeps into AI decision-making without intention. How you ask the question and what type of language you use to describe the issue is in itself indicative of the subconscious bias you carry. We tend to ask questions and frame problems through our own social and cultural lenses, while problems and issues to be solved tend to have their own cultural and social context. For instance, can a research consulting company based in London or Paris frame or define a problem for a community living in a city in Africa?

As we say, machines don't discriminate. It is the dataset that discriminates. And datasets by themselves do not discriminate. It is the way we handle datasets that can create a bias in AI decision-making. This is why it is important to train data scientists, and generally speaking, those who handle data in the AI space, on some basic ethical research principles so you can teach engineers to teach machines how to make better human decisions.

The Planning of Change: AI Management System Change Strategy

The last two requirements of the planning section (6.2 and 6.3) are devoted to the way you plan and implement your AIMS change strategy. A critical recommendation of ISO/IEC 42001 is to plan your AIMS objectives at different levels of the organization (enterprise, business unit, function, etc.) in a way that is aligned with the overall strategy, and consistent with the AI policy. Lower-level objectives, according to ISO/IEC 42001, need also to be monitored, updated, and communicated by organizational objectives so employees are aware of the linkage between these objectives and the overall business and AI objectives.

A good framework that helps you implement this section is the well-known Kotter model of change. As I have argued, AI, as a new technology with the power of a general-purpose technology, will impact the entire organization. Implementing an AIMS without paying attention to this particular dimension may affect the overall AI planning process and may jeopardize the effectiveness of the AIMS. When you adopt a consistent framework with a proven methodology of change, implementation becomes easier.

Next I will explain the process by which you should plan the AI change.

Create a Sense of Urgency

When implementing your AIMS, rallying those who will be in charge of the AI systems' implementation behind common objectives is a challenge. Because people are also doing other things, have their priorities, and are focused on their day-to-day KPIs, their dedication to the AIMS may not be fully secured. As argued by Kotter, injecting a sense of urgency motivates people to reset their priorities. The sense of urgency is a powerful marketing tool that helps you deal with complacency, especially with skeptics. Without a sense of urgency, AIMS defaults to a backseat position on people's busy schedules. So, make your AIMS change strategy a now-priority task.

Build the Guiding Team

The second step in Kotter's model is to build the team that will lead the change. When choosing people who lead the AIMS implementation, use the following suggestions:

- Identify employees available to meet as frequently as possible.
- Include a diverse background (different functions and professional background as well as cultural background).
- Appoint a senior leader with enough credibility and cloud to lead the change. Their role in setting the tone is of critical importance.

Get the Right Vision

In Kotter's model, getting the right vision means aligning people around the purpose of the AIMS. I have already explained in previous sections how you establish a vision according to ISO/IEC 42001. Here you are connecting people with that vision with emphasis on the *why* of the AIMS, and *how* it should be implemented.

Communicate for Buy-In

The opportunities and threats that the new AI age brings should be enough of a motivator to engage people around the AI system vision. However, because people may have a different perception of the technology, you need to work on a communication plan that helps you secure the buy-in of your internal stakeholders. Getting the buy-in is your first cognitive hurdle to rally the organization. Work with the marketing team to design a communication plan over a long period to explain the AIMS, and why you need to invest in it.

Empower Teams

The next step in Kotter's model is to empower those who will be in charge of the AIMS. There are many actions you can take to empower the team, like:

- Showing the opportunities, the AIMS will bring to the organization.
- Discuss the threats of not investing early on in an AIMS and how this will impact the organization.
- Allow for risk and failure during the implementation to enhance the psychological safety of teams in charge of the AIMS.
- Focus on collaboration and cross-functional teams with industry knowledge.
- Help your team see the outcome of their work by creating short-term wins to encourage and motivate them.

Perseverance

The last step in Kotter's model is perseverance. Kotter calls this step the "don't let up and make it stick" step. In this step, your executive presence, constant engagement, and motivation are of critical importance as they help employees stay focused on the AIMS implementation.

Chapter 8

Clause 7: Support

PURPOSE

The purpose of this chapter is to explain the AIMS support system. A focus on tangible and intangible resources, as well as the documentation required by the standard, helps the reader understand the overall requirements of the standard.

OVERVIEW

The chapter explains in detail the type of resources and support required by the ISO standard. It explains the role of the infrastructure, such as the computing power, storage capacity, and networking environment, and shows how a good investment in knowledge, soft skills, and attitude (KSA) enhances the functioning of the AI management system. The chapter also shows you how to create awareness and communication strategies to ensure that employees are informed about the AI management system and get their engagement and support during the implementation and certification process. A discussion of the documentation required by the standard and how to prepare for it is also presented with examples and samples.

DOI: 10.4324/9781003463979-10

BENEFITS

At the end of this chapter, you will be able to:

- Understand the overall support system required by the standard and how to prepare for it.
- Learn the importance of investing in the technology infrastructure to make your AIMS is a viable one.
- Learn the type of knowledge, skills, attitudes, and overall behaviors that need to be put in place for a good functioning of the AIMS.
- Develop an understanding of how communication and awareness impact the overall employees' engagement.
- Understand the type of documentation required for certification purposes.

Introduction

Section 7 of ISO/IEC 42001 addresses the support question. By support, I mean the infrastructure and all the actions that you must put in place to enhance the efficiency and effectiveness of your AIMS. The standard mandates five types of support:

- Resources: The tangible and intangible resources that you need to put in place for the AIMS to function. Things such as infrastructure, data, tools, computing power, networking, manpower, talents, etc. (Clause 7.1).
- Competence: The framework that identifies the knowledge, skills, and attitude (KSA) required of employees involved in the AIMS (Clause 7.2).
- Awareness: The campaigns that you engage in to inform and raise awareness of your stakeholders about the role and the importance of the AIMS, as well as the risk associated with your AIMS (Clause 7.3).
- Communication: The communication strategy you intend to put into place to explain and communicate with stakeholders about matters that relate to AIMS, including PR campaigns with external stakeholders to explain your AI product, AI strategy, and overall AI approach. Important also to add in the communication strategy is how you address scenarios of reputational risk (Clause 7.4).
- Documented information: The documentation, records, and results of studies that relate to the functioning of the AIMS. You will notice that

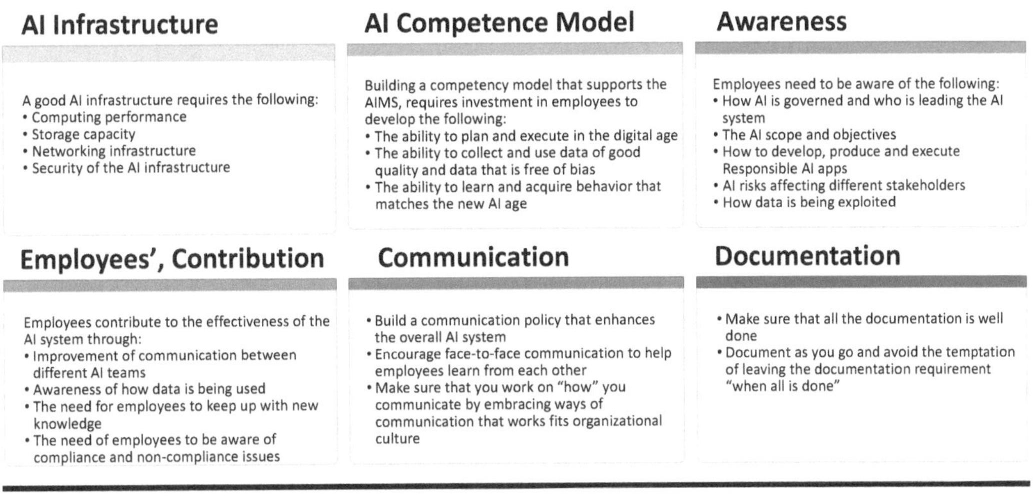

Figure 8.1 AI management system support: elements of an effective support system according to the ISO standard.

some of the documented information required is generic which you can find in other requirements standards such as ISO 9001:2015 Quality Management System, but some of these documents are specific to the AIMS.

Furthermore, Annex A provides more details on the type of documented information needed. Annex B provides helpful guidelines and explanations on how to prepare the documented information. Important to mention that the standard provides flexibility in the way you prepare the documents. In other words, the standard does not prescribe a specific template, beyond the generic requirement of any ISO document such as identification, dates, author, etc. You can prepare your documents in any way that fits your type of business, as long as it justifies and explains your rationale in documenting what is required.

In the next section, I will discuss the most critical support actions that you need to have in place to support the effectiveness of your AIMS. Figure 8.1 provides an overview of the support ecosystem needed to enhance your AIMS according to the ISO Standard.

Tangible Resources: The AI Infrastructure

As we move toward a more general intelligence system, relying on traditional IT infrastructure to manage AI systems will impact the effectiveness of your AIMS and may not allow your AI infrastructure to grow and develop. Relying on traditional IT infrastructure may even impede the growth and development of your AI systems. AI systems require more powerful computing machines, cloud and hybrid infrastructure, storage capacity, and large amounts of data that legacy IT systems may not be able to deliver. This infrastructure not only allows for the AI systems to be more effective in achieving their objectives, but it can also drive innovation as it helps talents using cloud and hybrid to make connections with other open-source systems, which leads to more creativity and more innovative products. Investing in a modern infrastructure that supports intelligent systems allows you also to gain independence from third-party providers and help you set the long-term foundations of an emerging ecosystem that will free your AI from IT legacy issues.

In his book, author Simon Thompson (2022) argues that, when it comes to managing large-scale ML projects, business leaders need to be aware of three types of requirements: Functional requirements, nonfunctional requirements, and system requirements. While the functional requirements refer to the performance of the model itself, the nonfunctional requirements and the system requirements, refer to the infrastructure needed to allow the ML projects and, more generally, the AI systems to achieve its performance goal. Author Terry Storra, from Leaseweb, a global IT infrastructure company, identifies four critical infrastructure assets that enhance the effectiveness of AI systems. These assets are the computing performance, the storage capacity, the networking infrastructure, and the security of the infrastructure.

Computing Performance

The performance of your computing power and how much calculation it makes raise essentially a hardware question. The ability of your AI systems to process large volumes of data, make complex calculations, and yet produce accurate predictions is intimately tied to the type of hardware used. The discussion to move from CPU to GPU, for instance, to upgrade your computing processing power is an important discussion to have. While not all companies can afford a GPU investment, especially if you are a medium or small enterprise, investing in a GPU should be considered as a long-term investment. As the proliferation of data accelerates, companies will rely more and more on data to make new products and services.

Storage Capacity

As we collect more data, the storage capacity becomes an issue. While some companies may not need a larger storage capacity, others may need to support their AI systems. As we know, ML algorithms make better decisions and provide more accurate predictions when they are exposed to larger sets of data. When an ML model has access to more and better data, its level of accuracy and overall sophistication improves tremendously, especially for complex ML algorithms that are deployed on real-world issues. So, the question of storage capacity is a critical question to ponder as it involves other issues, such as cost, security, scalability, and compatibility with existing infrastructure.

Furthermore, data annotation, the process of labeling and adding metadata to create a better context for the model, is intimately linked to storage capacity. According to the Zeitgeist AI readiness survey, a survey published in 2023 by Scale AI, organizations that invest in infrastructure create a better environment for data annotation, which itself leads to faster model production and more frequent reviews and updates of the model.

Networking Infrastructure

Communication between different deep-learning algorithms requires speed and reliability. AI systems may not deliver on their promise if the networking and communication infrastructure do not match the demand level. Of particular importance to the networking issue is the latency problem, and the time it takes for a system to respond to inputs, like in the case of speech recognition and self-driving cars. While investment in upgrading the networking infrastructure incurs more cost, in the long run, it dramatically improves AI system latency, a critical element for enhancing customer experience.

Security

The security of your AI system infrastructure is of paramount importance. Whether you decide to invest in CPU or GPU, or upgrade your networking communication infrastructure or not, securing the AI system infrastructure is a must. AI system handles sensitive data that, if breached or poisoned, will have legal implications and disastrous consequences on the business performance and the reputation of the organization.

Finally, to conclude the section on tangible resources, it is important to take a look at both Annex A and Annex B of the ISO/IEC 42001. These two annexes provide more details on the type of resources needed and how you should approach their implementation.

Let's take a look now at the intangible resources, knowledge, skills, and attitude needed to run the AIMS.

Intangible Resources: AI Competence Model

The second important clause of Section 7 addresses the issue of the competence model. While the standard does not require any specific competence model, it does suggest that for an AIMS to run effectively, a competence model that defines the type of knowledge, skills, and attitude (KSA) needs to be in place.

So, let's define what a competence model is, then discuss how you should build a competence model that is focused on AIMS.

What Is a Competence Model?

A competence model is a framework that helps an organization define the knowledge, skills, and attitudes (KSA) required for employees to perform. It is an inventory of skills and behaviors that the job description identifies as a requirement to perform on the job. A competence model identifies individual capabilities and answers the following basic question: "What type of knowledge, skills, and attitudes are required so my employees achieve their goals?"

To put in place a competence model that supports your AIMS, you need to work with your HR and Training and Development departments to identify the type of KSA needed for different employees involved with the AIMS. This is not an easy task to do as you need to pay attention to three challenges: The challenge of shifting job descriptions required in an emerging field, the issue of legacy facing the building of a new management system, and the question of interdisciplinarity needed to coordinate complex management systems such as AIMS.

- First, for an emerging field, such as AI, identifying the knowledge, skills, behaviors, and overall capabilities needed is not an easy thing to do as job requirements continue to shift. The knowledge complexity of the AI field and the interconnectedness of the AIMS with other knowledge domains make it difficult to come up with an agreement on what constitutes a competence model for an AIMS. The disruptions the workplace is going through today and the cultural change taking place in many industries require that you come up with an agile approach to defining the type of capabilities needed to run an AIMS. So, start by moving away from the traditional job description process to adopt a nimble approach that focuses on results and takes into consideration the general shifts affecting the workplace.

- Second, putting together an AIMS for the first time in an organization has its challenges. Building an AIMS on top of a legacy management system is a daunting challenge. Start small and focus on the core people doing the actual AI job. Then, expand from there to include all types of talents needed for the AIMS to operate, and how these talents interface with the new AIMS.
- Third, adopting an interdisciplinary approach to mapping out the individuals' competencies needed is a must. Because AI cuts across many fields and specializations, a comprehensive approach and a systemic view to building the skills and individual capabilities needed make your AIMS more effective.

AI-Focused Competence Model

A good starting point to build an AI-focused competence model is to start with a competence model that is comprehensive and takes into consideration the specificity of an AIMS.

A newly published model by the Broadband Commission for Sustainable Development provides a nice framework that defines in detail the type of knowledge, skills, and attitudes needed for an AI system. Although written for public service employees in general, I believe it is a good framework that can be applied to all types of organizations. It aims to help leadership think of an AI competence model more comprehensively and systematically when putting together frameworks for AI systems.

The Broadband Commission model (2022) identifies three competencies' domains, fourteen skills, five attitudes, and four proficiency levels that measure the performance of each competence domain. Let's go over the models and their competencies.

Competence Domain 1: Digital Planning and Design

The first competence domain of the model is the digital planning and design capability. This knowledge domain identifies concepts, principles, and practices that relate to the planning and design functions of digital transformation. For instance, technical proficiency, when combined with a strategic and anticipatory thinking approach, provides talents with the ability to plan and anticipate events that may affect the performance of an AIMS. An outside-in approach combined with a risk-based perspective on technology progress, on the other hand, helps you identify not only the market opportunities of your AIMS but also the risks associated with different AI applications. So, building this competence domain to run an AIMS is critical.

Furthermore, digital planning and design competence needs applied skills that allow employees to become more nimble. The Broadband Commission model identifies four sets of skills that, when combined with digital planning and design skills, will allow the AIMS to be more effective:

- System thinking: The ability to link different parts together (thinking wholistically).
- Problem identification: The ability to identify the problem and deal with ambiguity.
- Strategic foresight: The ability to project oneself into the future and think strategically.
- Agility: The ability to shift and pivot without losing focus.

Employees working on an AI project, who exhibit these set of skills, will be able to achieve better results. For instance, agile employees adapt better and respond faster to change requirements while problem identification skills allow teams to agree on the definition of the problem, which should allow for faster execution. System thinking skills help employees think in a more comprehensive and integrative way in addressing AI project planning and execution.

Model Competence Domain 2: Data Use and Governance

The second competence domain in the Broadband Commission model addresses the issue of data usage and data governance. This competence domain is critical to the functioning of an AIMS because it helps employees understand the process by which they turn raw data into exploitable insights while understanding the risk that is inherently built into the data. Data management, a field that used to be a vertical and obscure specialization for IT professionals, has become a horizontal function that impacts every aspect of the AIMS. Knowing how to deal with it not only helps employees leverage it in their decision-making process but also makes them aware of its harm to the user and the company. So, this is why your competence model should focus on data governance issues, so employees involved with the AIMS know how to collect data, how to label it and sample it, and how to maintain it and share it while fully compliant with privacy and personal information requirements throughout the AI project life cycle.

Skills that help you operationalize this competence domain, according to the Broadband Commission, are the following.

- Digital literacy: The ability to read, work, interact, and exploit a variety of digital media with ease to improve the performance of the organization.
- Data-driven decision-making: The ability to make decisions based on data trends and what the data suggests.
- Open data: The ability to take advantage of the open-source system and exploit data that is in the public realm to improve the organizational AIMS.
- Legal and ethical skills: Being aware of the laws, regulations, and policies that apply to AI technology, while being ethically responsible for AI usage, especially when it comes to data governance, privacy, and harm.
- AI fundamentals: How AI functions connect with other technologies and management systems, and the way AI has changed and will change the economy.

When the AI team is trained on these skills, its ability to take advantage of the many opportunities the AIMS offers is limitless. Your employees become better at assessing the true value of an AI project and can identify the vulnerability of data breaches in the system early on, which makes the AI project resilient to attacks and breaches.

Model Competence Domain 3: Digital Management and Execution

The Digital Management and Execution Competence domain, according to the Broadband Commission model, refers to the ability of the AI team to plan and execute an AI project. Beyond the planning and the design, the issue of the execution is of paramount importance to the success and failure of an AI project. In the details of the execution resides the difference between failure and success. So, a critical competence domain for an AIMS is to ensure that employees involved in AI projects are good at managing and executing projects. Attention to detail, being nimble in the execution, social perceptiveness (aware of other people's perception), time management and meeting deadlines, and transparency and courage to raise issues that may impact execution are the competencies that are needed for an AIMS. This competence domain gets enhanced with the following skills according to the Broadband Commission model.

- People Centricity: The ability to focus the attention on people. When dealing with customers it is customer-centricity and when interacting with employees it is employee-centricity.
- Iteration: The ability to interact, exchange information, and work with feedback.
- Agile execution: The flexibility in execution and being able to adjust deadlines and move things around during the execution.
- Agile Leadership: The ability to help people understand agility and train them how to acquire it through delegation and empowerment.

According to the model, these skills are critical to the AI project execution. Customer centricity, for instance, allows team members involved in managing AI projects to learn about project pain points. It helps them put customers' needs and wants at the core of the AI project. Iteration skills, on the other hand, allow for the team to integrate feedback from different stakeholders involved in the project into different stages of the AI project life cycle, which should impact the overall effectiveness of the AIMS.

Model Competence's Attitudes

In addition to the skills and knowledge domains required of the AIMS, the Broadband Commission model also defines the types of attitudes and behaviors that are suitable for an AI project. Attitudes and behaviors are key to the successful implementation of an AI project because they provide us with a window into the value system, the culture, and the inner thinking of people. Learning about how people perceive things and interpret the world around them, what motivates them, and what turns them off helps you design strategies that reinforce positive behavior and eliminate negative behavior.

Now that I have explained the three competence domains of the Broadband Commission model and the skills they require, let's turn now to the five attitudes the model provides.

Creativity

The ability of people to generate ideas, think critically, and solve problems using new and recycled ideas is critical to the effectiveness of an AIMS. This attitude is important because it helps your AIMS become more innovative in addressing new challenges and pain points.

Adaptability

The ability of people to adjust to new and changing circumstances, pivot to more critical issues, and respond faster to new requirements is an attitude that has become critical. Training people interacting with the AIMS on acquiring this attitude makes your AIMS more agile and nimble, especially during uncertainty.

Experimentation

For an emerging technology, people with an experimental mindset add tremendous value to AIMS. As a nascent technology, the potential of AI is tremendous. People who try things out and experiment with failure bring additional and more strength to the performance of the AIMS.

Curiosity

This attitude is linked to experimentation. Curiosity is what fuels the creativity and innovation of people involved with AIMS.

Trust

As an attitude, trust is critical to experimentation. Trusting your employees to bring new ideas, play, and experiment with those ideas helps your AIMS become more innovative. If you remove trust and replace it with compliance, people default to a traditional job description that may impede innovation and creativity.

Finally, for the competence model to work, it has to be clear and communicated. As well described by a PwC study, a successful competence model has to have "4Cs."

- Clarity of what is required to succeed.
- Consistency in how you measure the skills needed.
- Connecting skills and competencies to the organizational strategy.
- Continuity in the approach and the message being taught.

Awareness (Section 7.3)

For a technology that will transform the organization, communication of its impact and awareness of its effects are key. Communicating the objectives of the AIMS and explaining to different stakeholders the scope and the impact of AI technology is a critical dimension of an effective AIMS. Leadership needs to understand this dimension as well as its implications on the overall organization's performance and liability. This is why ISO/IEC 42001 (Section 7.3 and Section 7) requires that organizations design an awareness campaign and communication strategies to accompany the AIMS implementation.

Three requirements relate to the awareness clause.

- Organizations must inform and make employees aware of the AI policy.
- Organizations must show how employees contribute to the overall effectiveness of the AIMS.
- Leadership must make employees aware of the risk of not conforming to the AI Management system requirements.

Let's go over these three requirements:

All Employees Need to Be Aware of the AI Policy

This is a relatively strong requirement as the leadership is expected to engage in an awareness campaign to explain the AI policy of the organization. The standard does not specify what exactly needs to be communicated to employees about the AI policy, but I would like to suggest that for your awareness campaign to be engaging, you need to communicate at least the following elements:

Governance and Leadership

Employees need to know who is in charge of AIMS. They need to be aware of different roles and responsibilities within the AIMS, as well as the governance mechanism by which leadership exercises oversight. Employees need also to know how to reach different leadership levels in charge of the AI governance when they need to report something about the AIMS.

AI Scope and Objectives

An AIMS exists to achieve business goals and strategic objectives. Employees need to know the objectives of AIMS, its scope, and why the organization invests in it. When employees are aware of leadership thinking about the AIMS, their commitment and engagement increase. Being open about the goals and the objective also impacts positively organizational trust.

Use of Responsible AI

This is another critical dimension of the AI Management system that needs to be well communicated to employees. As AI becomes more mainstream technology and as certain AI applications become more controversial, leadership needs to take a proactive approach to clarify how the organization intends to use AI in a responsible AI way. Responsible AI simply means that the organization subscribes to the laws and regulations that guide the use of AI, as well as the ethical principles of do not harm, fairness and nondiscrimination, privacy protection, and explainability.

AI Risks

Another equally critical aspect of the AIMS that employees need to be aware of is the issue of risk. In many ways, this is a challenge for leadership. A good approach to deal with this challenge is to come up with a risk classification approach that identifies risks from high to low, and how the organization intends to deal with different levels of risks. Employees, especially those involved directly with the AIMS, need to be aware of the tools and the methodology the organization adopts to measure and assess risk, so these risks are integrated throughout the AI project life cycle.

Data Usage

Finally, the issue of data. A critically important aspect of the AIMS that employees need to be aware of is the company's policy on data and how employees should use data. This is the responsibility of leadership and management at different levels to make sure that their employees understand the inherent risk to data and the company's policy on data, especially, who and when to access data, how to process it, and with whom to share it.

How Employees Contribute to a Better Improved AI Management System

The second requirement of the awareness clause is for leadership to ensure that employees are aware of their contribution to an effective AIMS. For an AIMS to function effectively and efficiently, it requires many moving parts. Many of these moving parts that optimize the effectiveness of the AIMS are not centralized in one unit, like in the case of traditional IT. As technology seeps into all facets of the organization, the challenge is to help employees understand that their work, even if it does not appear to be tech-related work, does in many ways impact the overall AI ecosystem. For sales, customer service, marketing, and HR teams who may think of themselves as not traditionally IT teams, their impact and their contribution to the effectiveness of AIMS are critical, especially with upstream teams dealing with data.

Following are a few suggestions that help different teams interacting with the AIMS improve the performance of the system.

Communication between Different AI Teams

An AIMS requires input from a variety of teams: Engineers, sales, marketing, HR, and so on. Depending on the context of the AIMS, seamless communication between these different teams optimizes the effectiveness of the AIMS and creates a better integrated AIMS.

The Use of Data

Because data comes from everywhere, teams engaged in managing and processing data management need to be aware of their upstream role in the overall health of the data ecosystem. When upstream teams provide low-quality data, this may affect the overall AIMS. The leadership role in raising the awareness level of employees manipulating data is critical.

The Need to Reskills and Upskill

As a nascent technology, AI requirements keep changing. AI knowledge itself keeps shifting, so helping employees gain the proper knowledge and skills and helping them translate that knowledge into practical processes allows for the AIMS to be better maintained and up-to-date, which should help the overall performance of the AIMS.

Noncompliance Issues of the AI Management System

Finally, the last requirement of the awareness section is to ensure that employees are aware of the compliance risk. Compliance risks are a serious threat to the organization and the AIMS. As AI technology becomes a societal issue and more ubiquitous, pressure for accountability and transparency in how the organization complies with ethics, laws, regulations, and standards increases. The leadership role is to explain and show employees the compliance process and how the organization intends to subscribe to it. Here are a few steps that could help your employees engage with compliance:

- Encourage your teams involved with the AI project to engage with the legal department early on during the AI project life cycle to flush out compliance issues.
- Search AI legislations and regulations in your country and other countries where you operate.
- Explain to your employees how your AIMS should comply with these regulations and help them understand the dos and don'ts of your AI applications.
- As AI regulation and policies keep shifting, create a watch list of AI-related legislations to monitor new and changing requirements in the world that may affect your AIMS. Countries' policies also are shifting, so it is critical to keep an eye on new changes in policies and requirements for AI systems, especially in jurisdictions where you operate.
- Search for international standards that guide organizations on compliance issues (ISO/IEC standards, NIST framework, European standards, etc.) and get certified on the main ones. Of critical importance, is the new set of standards that will be developed by the European Committee for Standardization (CEN) and the European Committee for Electrotechnical Standardization (CENELEC) to operationalize the new EU AI Act.

Communication (Clause 7.4)

Depending on the context and the economic sector in which you operate, Section 7.4, Communication, provides you with the flexibility to define the best ways to communicate about your AIMS. As long as your AI communication strategy identifies the target audience (who will benefit from this communication), the frequency of communication (how often and when you communicate), the medium of communication, and the objective of the communication (how and why you communicate), the standard provides you with enough flexibility to set the AI communication strategy that fits better your context.

Following are some best practices of a good AI communication policy:

Encourage Face-to-Face Communication

While emails, bullet boards, and other electronic media are a good way to explain a new policy and raise people's awareness of the policy, communicating a new and complex system, such as AIMS, requires direct communication. To employees, direct communication has an enormous benefit. It shows that leadership cares about AI technology and considers it as a high-priority business issue. It also helps employees understand leadership inner thinking, so they can connect better on such strategic issues.

To leadership, direct communication helps them set the tone and build the trust needed during large-scale transformations. Direct communication also helps leadership understand granular implementation challenges and how to address them.

The Medium Is the Message

People don't care about what you say but how you say it. When leadership engages personally in communicating about the AIMS, employees sense better the urgency and the interest leadership devote to AI. Personal engagement of leadership in the AIMS sends a far more important message about the AIMS. It shows that leadership pays careful attention to the investment the organization makes in building the AIMS and is willing to take actions to improve, support, and reward those engaged in making the system a better one.

Create Policy Champions

A new change always requires a change champion. AIMS champions are crucial to the implementation of the new system. Champions are critical because they are employees who care about the success of the AIMS and are passionate about the new change the AIMS brings to the organization. When change champions are in a leadership position, they lead by example and become role models. Champions also help alleviate resistance and engage skeptics in a discussion about the benefits of AI investments.

Documented Information

Clause 7.5 on documented information of the AIMS is generic. You can find the same clause in many other standards, such as ISO 9001: 2015 Quality Management System, ISO 45001: 2018 Occupational Health and Safety, and ISO 14001: 2015 Environmental Management System.

In this section, I will first provide some general guidelines on how to handle AIMS documentation. Then, I will identify all the actions that must be documented. A simple quick keyword search of the standard will bring up all the required documentation you need. So, let's start with a few general remarks.

- First, important to mention that ISO/IEC 42001 requires two types of documented information. The documented information required by the standard itself, and the documented information that you deem necessary for an effective functioning of an AIMS. Organizations differ by size, sector, activities, and complexity level so it makes sense that documented information on the AIMS may differ from one organization to another.
- Second, when you create and update documentation, documents need to be identified by title, date, author, and reference number. These are general requirements of all ISO management systems. Furthermore, documents can be in any format that suits the purpose of the activity. Documents are also needed to show the authority that reviews and approves the documented information.
- Finally, documented information that relates to the AIMS must be protected and available. This is a requirement that is meant to create

control over access to the documentation, and the maintenance of the documentation itself against damage. The same controls apply to the documented information that is external to the organization and deemed necessary for the effectiveness of the AIMS.

Documented Information Required: What Needs to Be Documented

For certification purposes, ISO/IEC 42001 requires the following documents:

1. The scope and the activities of the AIMS, its boundaries, and its applicability, in addition to the overall context of the organization (Clause 4.3)
2. AI policy must be available and documented (Clause 5.5)
3. AI risk and AI opportunities: This is a critical document to have. There are many controls of the risk analysis, so providing a registrar of all the risks is critical (Clause 6.1)
4. How the organization conducts its risk assessment (Clause 6.1.2)
5. Three types of documentation relate to the treatment of risk. Justification of why an organization excluded certain types of control. Justification of why the organization established its controls. And, finally, how and in what way the organization intends to treat risks.
6. The process and results of the impact assessment must be documented. The standard provides the option to the organization to share this documentation with interested parties (as defined by the organization) (Clause 6.1.4)
7. AI objectives at different levels of the organization and how the organization plans to achieve them (Clause 6.2)
8. The skills, knowledge, and competence of the people engaged in the AIMS (Clause 7.2)
9. Documentation of how the planning and control process works. This type of documentation is required only to make sure that things go as planned.
10. Documented information on the results of the three activities that relate to risk: Risk assessment, risk treatment, and impact assessment (Clauses 8.2, 8.3, 8.4)

11. Documented information on the monitoring and measurement activities of the AIMS (Clause 9.1)
12. Documentation of the audit program as well as the audit results (Clause 9.2.2)
13. The results of the management review of the AIMS (Clause 9.33)

Chapter 9

Clause 8: Operation

PURPOSE

The purpose of this chapter is to explain how to plan, manage, and control the process of AIMS creation. The chapter outlines the required controls for AIMS performance and explains the process of putting in place the required controls for compliance. It also provides a framework to manage the AI project life cycle.

OVERVIEW

The chapter also provides guidelines and examples on how to implement the different controls required by the ISO standard. The reader will find a framework on how to manage the AI project life cycle and the type of controls needed to be in place for the AI project to meet the ISO certification requirements. The goal is to create controls that enhance the implementation of the AI management system and provide guardrails for better functioning of the AI management system. The chapter also explains the steps that relate to the design, development, and deployment of the project life cycle as well as the process grouping needed for each step to be compliant with ISO requirements.

BENEFITS

At the end of this chapter, you will be able to:

DOI: 10.4324/9781003463979-11

- Learn how to put controls in place to ensure a viable performance of the AIMS.
- Understand how to set up control that makes your AIMS compliant with ISO/IC requirements.
- Identify the issues that may affect the certification process.
- Learn how to integrate controls into the different steps of the AI project life cycle.

Introduction

Clause 8 is about operational control. Operational control helps you manage, administer, and monitor different processes and procedures that allow your AIMS to function effectively and to meet ISO standard requirements. Clause 8 is intimately linked to Clause 6 (Planning), so it is important to read it in conjunction with the Clause 6 requirements, which focuses primarily on risk assessment, risk treatment, and impact assessment.

In simple terms, Clause 8 (operational control) answers the following simple question: How do I make sure that the processes and procedures put in place function as they were intended to? How do I monitor and control these processes?

ISO/IEC 42001 requires three critical steps for operational control:

- Define the required controls needed for the functioning of the AIMS.
- Establish a system to monitor, control, and provide corrective actions when needed.
- Document the planning and control process to enhance confidence in the system.

To discuss the required controls, I will follow an AI Project Life Cycle approach. I will explain each stage of the AI project life cycle and discuss the controls required by the standard.

AI Project Life Cycle

The purpose of an AI project is to solve a problem. Whether it is about automating mundane processes and repetitive tasks or creating completely new business opportunities, the AI project's goal is to take advantage of ML technology to improve the organizational decision-making process and impact the bottom line. A challenge, however, that organizations face in managing AI projects is the complexity of the AI project.

An AI project differs from a traditional IT management project in many ways. An AI project is more complex, is highly uncertain, has higher risks and costs, and has many moving parts that may or may not fully align. So, a critical step in managing an AI project is to have a clear understanding of the framework that you need to apply to manage the entire AI life cycle of the project.

There are many frameworks you can choose from to put together a process to manage, execute, and improve an AI project life cycle. ISO/IEC 5338:2023 Artificial intelligence: AI system life cycle processes is one of them. But a simple framework that I would like to suggest is the US GSA AI Guide for Government. Made of three basic steps – Design, Develop, and Deploy – this framework is quite simple to follow and implement (see Figure 9.1).

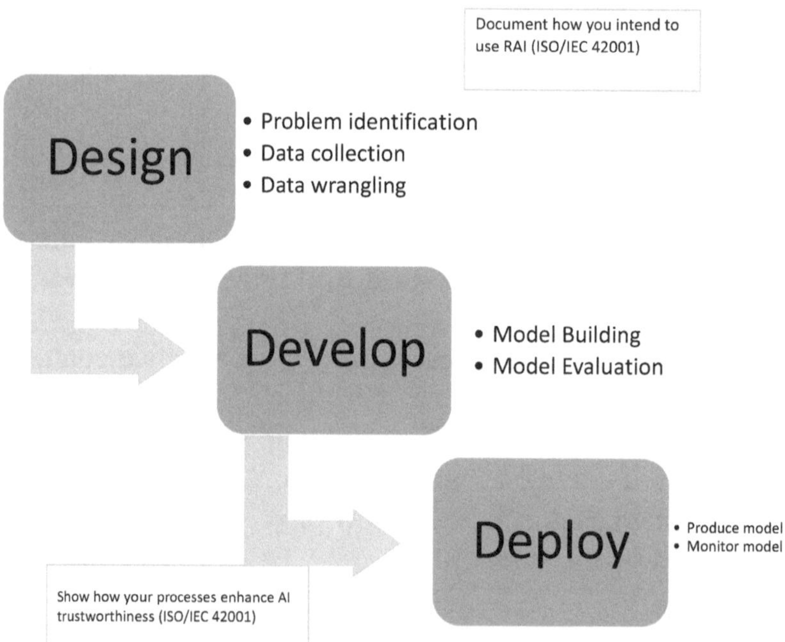

Figure 9.1 AI project life cycle.

Adapted from AI Guide for Government: A living and evolving guide to the application of Artificial Intelligence for the U.S. federal government. GSA. May 2023. https://coe.gsa.gov/ai-guide-for-government/

Next I will discuss each step of the model and provide the controls ISO/IEC 42001 requires for each step.

Design Phase: Process Grouping 1

Three steps in the design phase need to be followed:

- Identify the problem the AI project needs to solve.
- Gather the data needed for the AI project.
- Clean and prepare the data.

Identify the Problem

The first step is to identify the problem that your AI project needs to solve. A thorough discussion must take place between the project owner and the AI team to understand the nature of the problem, the pain points, and the issue the ML project is trying to solve. In this stage, the focus is on understanding the business problem and the scope of the project. In this step you need to do the following:

Select the Idea

During the process of problem definition, you will find yourself overwhelmed with ideas that may compete to solve the problem. A careful selection of the idea that fits better the definition of the problem is critical to developing a good definition of the problem.

Understand the Context of the Organization

You will not be able to define the goal and select the problem statement unless you have analyzed the context of the organization. Context analysis of the organization and the problem requires a good knowledge of the industry, the organization, the competition, and how the AI project is supposed to add value. Given the complexity of the context in which your organization operates, a thorough understanding of the context entails an understanding of the following:

- The mission of the organization.
- How the AI project adds value

- The nature of the relationship the organization has with its wider network of stakeholders.
- The competitive landscape of the organization.
- The market forces impacting the organization.

Conduct a Literature Review

A literature review helps you understand how other organizations solved the same problem or a problem similar to yours. The benefit of the literature review is that it provides you with a scientific perspective on how and why you need that particular technology to solve that particular problem. It also helps you understand how the research community feels about the type of ML technology you plan to apply.

Frame the Question

After you have defined the problem, analyzed the context, and reviewed the major research on the subject, it is time now to frame the question. The framing of the question is critical because this is what will guide engineers and data science throughout the AI project life cycle. The framing of the question becomes a road map that helps the team prioritize issues and make decisions as they progress. In the framing, you can:

- State the question.
- Define the problem in a clear and simple language.
- Define the desired outcome.
- Set initial KPIs and milestones.
- Identify the resources needed.

ISO/IEC 42001 Requirements for Process Grouping 1 (Design)

Responsible AI

In the design phase, you need to document how you intend to identify and develop responsible AI objectives that apply to your AIMS and how you plan to integrate them. This is a critical requirement because it spells out your philosophy about responsible AI and how you intend to implement it throughout the AIMS. Figure 9.2 provides a process grouping approach to the AI project life cycle.

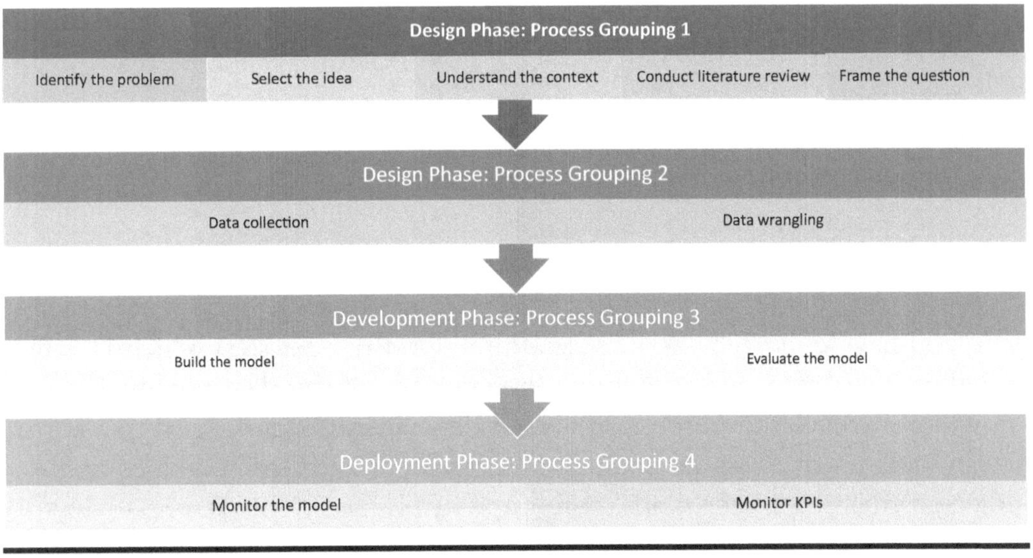

Figure 9.2 Adapted from GSA 2023. AI Guide for Government: A living and evolving guide to the application of Artificial Intelligence for the U.S. federal government. Retrieved from https://coe.gsa.gov/ai-guide-for-government/

Trustworthy AI

You need to explain and document how you intend to develop and design processes that enhance the trustworthiness of your AIMS. In other words, explain how and by what means you will trust your processes to develop an AI project that is trustworthy. Both of these requirements are listed in the Annex A of the standard (A.6.1.2 and A.6.1.3).

Design Phase: Process Grouping 2

This is a critical stage. This is where the technical work starts and where the issues that impact the performance of your model start to seep into the project. Investing more time and resources to get this step right, will save many hurdles, and will help your project avoid all future problems. As we say, it is the data that will make or break your AI project.

Data Collection

The first set of questions that you need to discuss are the following:

- **Data collection methods**: How do you collect data? There are many methods like surveys, experiments, focus groups, polling, etc.
- **Data sources**: Where does the data come from? Is it proprietarily owned by your organization or public? And how did you get it? Is it

primary data, data that has been collected by you directly? Or secondary data, data that has been collected by someone else and made available to you?
- **Size of the data** and how much of it you have and how much of it you need to run your project.

Data Wrangling

Once you have collected the data, you move to the next step which is data wrangling. Here you start looking at the specifics of the data quality. Data wrangling involves the process by which data scientists clean and transform the raw data to avoid issues such as missing records, extreme values and outliers, formatting issues, and, more importantly, bias. This is a relatively technical step that requires data scientists who understand well the issues that surround data collection and data preparation methods.

Data wrangling involves many steps.

- Cleaning the raw data that may come in different formats.
- Cleaning the data to fix errors and remove inconsistencies and outliers.
- Data conversion: Once you clean the data you are ready to transform it into a suitable model.

ISO/IEC 42001 Requirements for Process Grouping 2 (Design)

Important to mention that for this stage, the ISO/IEC 42001 requires many controls. It is critical that you look at the standard and Annex A to make sure that the methodology for data collection and data wrangling meets the standard requirements.

Data Quality

- How do you intend to use the data to enhance your AIMS? (Standard Annex A.7.2)
- How do you get the data, and how do you select your datasets? (Standard Annex A.7.3)
- How do you control the quality of the data, and what processes are used to make sure that the data meets your quality assurance requirements? (Standard Annex A.7.4)

- Where does the data come from? (Standard Annex A.7.5)
- What is the process by which you prepare the data to build the model? (Standard Annex A.7.6).

Data Resources

The identification of data sources is a requirement and is part of the standard controls. ISO/IEC 42001 requires that you identify the data sources of the AI project: What is it, how did you get it, how do you plan to curate it, and how large is it? (See Annex A. 4.3)

Development Phase: Process Grouping 3

Once you have developed a certain level of confidence in your data quality and checked that it meets your quality standard requirements, as well as the ISO/IEC 42001:2023 requirements, it is time to move to the next step according to the US GSA Guide. In this stage, your goal is to build the model, test it, and experiment with it iteratively so you can fine-tune the model as you go. There are two steps in this model development phase. Building the model and evaluating it.

Build the Model

In this step, the focus is to:

- Define the features of the model.
- Identify the parameters that best represent the model.
- Work with the domain experts to make sure that the model captures well the issues you are trying to solve.
- Use benchmark for evaluation of the initial model.
- Start your testing, training, and retraining based on the results you get so you can fine-tune the model.
- Keep in mind that no model achieves results the first time, so retraining the model is a must-step to improve the model.

Evaluate the Model

The evaluation step is critical as it allows you to test and evaluate the best model that meets your requirements. Testing is done usually on new data and your goal is to monitor how and in what way the model behaves with new data. The test and evaluation also allow you to identify the risks and the issues that your model may raise.

ISO/IEC 42001 Requirements for Process Grouping 3

The ISO/IEC 42001 Standard requires that the tooling resources (programming languages, libraries, software kits, etc.) of the AI project must be documented. So, make sure that all the tools you use to build and develop the model are included in the documented information. See Annex A.4.4.

Deployment Phase: Process Grouping 4

Once your model achieves the acceptable performance level you set it initially, you need to move to the deployment step. This step is also called the production step. In the deployment step, the model should be able to take new data that was not part of the initial training set. Decide whether you want to deploy the project on-premises, in the cloud, or as a combination of both. It is also important to start monitoring the issues of cost, scalability, security, and accessibility and whether you are aligned with the initial estimate of the project.

Monitor Model Behavior

In this stage, careful monitoring of the model behavior is important as the model's behavior may change over time, or drift from its initial performance goals. One of the best practices is to set up an alert mechanism to start tracking performance issues. It is also critical that you keep retraining the model to help the model improve over time.

Monitor KPIs

In this stage, it is also critical that you pay attention to the initial KPIs you set and whether the model meets those KPIs. If the model is not meeting your initial KPIs then you need to either take a second look at the model or reassess your KPIs.

ISO/IEC 42001 Requirements for Process Grouping 4

The standard ISO/IEC 42001 requires that systems and computing resources involved in the AI project should be part of the documented information, so make sure you are documenting these resources as you go so you don't have to redo it during the certification process.

Finally, it is important to mention that Annex A of the standard lists many requirements that must be met during the management of the AI life cycle. I suggest that you check the list of controls in the annex to decide which one applies to your organization, and which you can substitute by your controls. Most of these requirements are documentation of the processes of the AI project life cycle, so make sure you keep all the records as you progress from one step to another.

TOOLBOX 22

ETHICAL PERSPECTIVE ON DATA MANAGEMENT AND DATA SCIENTIST OCCUPATION

The role of data scientists in ensuring that the AI system is free of bias and discrimination is crucial. As a new field, data science is at the intersection of many fields of study such as statistics, math, and computer science. It has a powerful impact on people's lives, so data science must have the ethical tools that allow them to understand how their system may affect people's life.

The Institute and Faculties on Actuaries (IFoA) and the Royal Statistics Society (RSS) in the UK developed an excellent guide that can serve as a framework for data scientists and help you address ethical questions as you go through the Design Develop Deploy model of data management. The guide raises some interesting questions and challenges facing engineers, managers, and scientists working in the field of data.

To help data scientists tackle these issues and challenges, the guide provides a framework made of five principles. These principles should guide data scientists throughout the whole data design, development, and deployment model.

1. **Seek to enhance the value of data science**. Because data in today's world has a huge impact on people's lives, data scientists need to

embrace an ethical framework and a mindset that helps them think of fairness and well-being as guiding principles when they use data. Data scientists, and those involved more generally in the process of data management, need to embrace a "public good" philosophy to the issues of economic empowerment, employment, education, environment, equity and inclusion, fairness, health, and hunger.

So, to enhance the value of data for society, the guide provides the following steps for data scientists:

- Ask how the model impacts stakeholders, in what ways, and how the issues of employment, education, and environment, for instance, can be impacted by the type of data you use to train your algorithm models.
- Identify the stakeholders who will be impacted by the model, their social and economic status, and whether the data truly is representative of their concerns.
- Ask if the algorithm and the data that fuels it can create potential harm. Ask about the types of bias and error your data may have and how this bias may impact stakeholders.

2. **Avoid harm**. The role of data in creating harm is real. Issues of privacy, personal information, damage to reputation, and long-term psychological effects on people and communities are real and have all been well documented. To avoid or minimize harm, the guide suggests the following:
   - Identify the sources of data: Where does the data come from, how has it been collected, and how do you intend to use it?
   - Have you obtained the legal consent to use data, how, and in what ways? For instance, do subjects know about how their data is being used?
   - Have you identified some best practices to handle the data with care? How to secure, store, use, and share personal information throughout the data project life cycle?
3. **Applying and monitoring professional competence**. The role of data scientists and, professionals involved in the handling of data, in general, is critical. The competence of the team, its skills, and its commitment to the highest ethical standard are critical factors in producing good and fair algorithms. As a professional involved in data manipulation, the guide suggests that you:

- Invest time to understand the legal framework, the ethical principles as well as the complexity and the nuances that surround the issues your algorithm model is tackling.
- Convey that understanding to decision-makers and ensure that those in charge of the final approval know the risks associated with different steps of the data manipulation.
- Apply the correct statistical method and the correct algorithm model to address the issues at stake, while still monitoring the model performance and results.

4. **Seek to preserve or increase trustworthiness**. The issue of trustworthiness in data is a critical challenge. People will not engage with your AI applications if these applications are not deemed trustworthy. The public backlash, when it happens against a product or service, impacts directly the issue of trust. A good practice, the guide recommends, is to work with the communities that will be impacted by the data you use to create trust and enhance the trustworthiness of the system. The guide suggests the following:
   - Engage and communicate with ethical and regulatory bodies to explain how you collect the data, what your data model does, and how you come up with a model.
   - Practice simplicity in communication and explanation. Use data models that are simpler for the public to understand and be proactive in explaining how they may impact them.
   - Practice transparency in building your model. People disengage with opaque systems and engage with transparent ones because they understand how they work and how it impacts them.
5. **Maintain accountability and oversight**. The fifth principle in the guide of the Royal Statistics Society and the Institute and Faculty of Actuaries is to ensure human oversight and accountability by showing who is in charge of the model, and how those in charge of the model can be held accountable if something goes wrong. Following are some of the suggestions the guide provides:
   - Establish a governance model to plan decide, review, and approve different stages of the data collection steps.
   - Establish and activate the role of the Ethics Committee in checking different data management stages.

- Engage and communicate with the public about how your oversight and government system work.

Finally, the guide provides a nice checklist that you can use in the different steps of your data management process, design, development, and deployment. You can find more information about the guides and ways of applying them at www.rss.org.uk and www.actuaries.org.uk

Chapter 10

Clause 9: Performance Evaluation

PURPOSE

The purpose of the chapter is to explain the AIMS performance evaluation requirements and outline the AIMS audit and process. It also provides examples of metrics used to assess the AI system performance.

OVERVIEW

The chapter defines how to conduct measurement and gives examples of metrics used to measure the performance of the AI management system.

The reader will find frameworks, tools, metrics, and KPIs that can be used during the assessment process. The chapter also describes and explains the difference between the two types of assessment involved in the AI management system auditing process: The governance, or the sociotechnical assessment, and the technical or algorithmic audit process, which focuses on the behavior of the AI algorithm. Finally, the chapter defines the role of management review as a requirement of the ISO certification process.

DOI: 10.4324/9781003463979-12

BENEFITS

At the end of this chapter, you will be able to:

- Learn how to conduct the AIMS assessment process.
- Understand the difference between sociotechnical assessment and algorithmic audit.
- Understand the assessment criteria required by the ISO/IEC standard.
- Develop a framework to measure AIMS performance for auditing purposes.
- Learn the role of management review during the auditing process.

Introduction

The success of any management system depends a great deal on the evaluation process of its performance. Management systems, complicated as they are, as they tackle many interrelated parts, need to be assessed and evaluated frequently to identify their shortcomings, spell out their weaknesses, and design strategies to improve them.

Performance evaluation is critical because it allows leadership to realign resources, objectives, and missions to achieve the goals. But for the AIMS, performance evaluation takes on a new meaning in the sense that not only does it provide an overall assessment of the system, but also it helps audit the algorithm and the data that fuels the AIMS. Algorithm audit, as a relatively new and emerging field, helps management address issues of bias and discrimination, as well as the question of whether the model is behaving according to the goals set by the organization. This is why I believe that when we address the question of AIMS evaluation performance, we need to address two essential, and separate, questions.

- First, whether, and overall, the system works in a coherent way creating synergies that maximize the collective potential of the different parts involved in it.
- Second, whether the algorithm and the data that fuels it are compliant with the laws and regulations, as well as the ethical framework of the company.

These are two different, and separate, assessments that both need to take place in the organization. While the first type of assessment can be conducted by assessors familiar with ISO management system frameworks, like ISO 9001 lead auditors, the second type of assessment is relatively complex and requires a technical background. There is a new standard under development, as of the writing of this book, ISO/IEC CD 42006 Requirements for Bodies Providing Audit and Certification of Artificial Intelligence Management Systems, that will define the audit and certification process as well as the skill set required of auditors. This new underdevelopment standard adds specific conformity assessment requirements to bodies providing audit and certification (ISO/IEC 17021-1:2015 Conformity Assessment Requirements for Bodies Providing Audit and Certification of Management Systems).

In this section, I will address both questions to show you the general process of AIMS evaluation. So, let's start with the standard requirements.

AI Management System Evaluation and Assessment Requirements

To evaluate the performance of your AIMS, ISO/IEC 42001 requires three important steps.

- Design a framework to assess, monitor, and measure the performance of the AIMS.
- Set up an audit program to assess whether the AIMS complies with your own organization's requirements, as well as with the requirements of the standard ISO/IEC 42001.
- Review the evaluation and assessment output to decide about ways and means of improving the AIMS.

Let's go over these three actions and provide some best practices:

AI Management System Assessment and Audit

The purpose of an evaluation system is to generate enough data that provide evidence of whether the system works as it was intended. The objective is to make evidence-based decisions to improve or change the system. Especially true for complex systems such as AIMS, the evaluation requires a technical review, as well as a sociotechnical evaluation that addresses the

Criteria	Question	Yes	No	NI
Leadership	How and to what extent leadership is engaged in the AIMS?			
Documentation	Are all the documentations required by the standard in place? Examples of documentation could be AI policy, AI frameworks, AI goals and objectives...etc.			
Supporting policies	Are the relevant policies that relate, either directly or indirectly, to the AIMS in place? For instance, HR and talent policies, procurement policy, and other policies that support the overall AI governance system.			
Roles & responsibilities	Have we identified and defined the roles and responsibilities of different actors involved in AIMS?			
Governing practices	Are the governing practices that relate to AIMS well understood and well defined? Do people understand them in a way that allows them to accomplish their mission?			
Governing bodies	Do governing bodies understand the risk of AI systems and are they aware of AI applications being developed?			
Leadership understanding of the Risk	Do leaders understand the AI risk? Have they been briefed on the type of AI risks we may run into? Have they cleared the impact assessment study?			

Figure 10.1 AIMS governance assessment checklist.

quality of the governance model. So, when deciding about the evaluation, you need to think in a more comprehensive way to address three questions. An assessment checklist is provided in Figure 10.1.

- What do I need to evaluate? (The scope of the evaluation)
- How do I evaluate? (The tools and methods of evaluation)
- Who will conduct the evaluation? (The skills of people engaged in the evaluation)

The Scope of the Performance Evaluation

Depending on the size of the organization and the specific sector in which you operate, as well as prior evaluation of the system and the AI maturity level of the organization, the performance evaluation can address the whole AIMS or only part of it. According to the standard ISO/IEC 42001, the organization decides what to evaluate, the whole system or only part of it.

Because ISO/IEC 42001 is a new standard, my recommendation is that you conduct a full evaluation. You may want to start with the technical side of the AIMS, like data, model training, and algorithm behavior, while another parallel team works on the governance issue, such as leadership, AI policy, AI strategy, context analysis, and risk management of the AIMS. Breaking down the assessment process in this way helps you not to overwhelm the assessment process while allowing the organization to digest the assessment results and make the necessary changes before certification.

Assessment Criteria, Metrics, KPIs

Once you identify the scope of the evaluation, you need to address the question of assessment criteria, metrics, and KPIs. There are two steps to the evaluation and measurement of an AIMS: The evaluation and measurement of the AI governance system as a whole, and the evaluation and measurement of the machine learning system. These two steps are linked, and they need both to be measured and evaluated.

The first step is to evaluate the governance of the AIMS.

Here you can use many criteria. Peter Soendergaard, from the Soendergaard Group, suggests five important KPIs:

- Data: Measuring the quality of data and the process by which it is handled is an important KPI to start with in the evaluation process.
- Security: How is the data handled, and how is the AI system protected from cyberattacks, such as evasion attacks, poison attacks, and extraction attacks, as well as other forms of cyberattacks?
- Cost Value: According to Soendergaard, this is a critical measure of the AI government system that helps you identify the real value of the investment and its ROI.
- Bias: Identifying and measuring bias in the data and the training model is critical as it helps you address legal and ethical issues.
- Accountability: This KPI helps you identify the people using the system and deciding about contribution to the system.
- Audit: Auditing the algorithm frequently to measure its effectiveness and efficiency, as well as identify ethical issues such as bias and discrimination.

You can also use the following table to assess the overall governance of the AIMS. These questions are derived from the ISO/IEC 42001 requirements.

The second step in the evaluation of the ML program is the evaluation of the ML model itself. Here are some of the ML metrics you can use. These are the criteria suggested by Jerald Murphy (2023):

- Mean Squared Error (MSE): a technique that identifies outlier results.
- Meantime to Repair (MTTR): the time it takes to fix things from first contact. This is a well-known KPI in IT.
- First Contact Resolution Rate (FCRR): A classical IT measurement technique that helps you identify the issues that get resolved by support service without further investigation.
- Error analysis: Analyze the prediction errors in the ML model to investigate the root cause.
- Data and concept drifting: A common ML error, this KPI helps you keep the ML model on target by monitoring data distribution and changes in data input patterns.

Large Language Models Audit

The democratization of foundation models through large language models (LLMs) and the many applications that have become a world sensation recently brought to the conversation table a critical question about the audit of these models. If your AI system is part of a large foundation model ecosystem, using applications produced by another provider, then the question of algorithm transparency and algorithm audit is a critical question to address.

While ISO/IEC 42001 does not address the issue of large language models *per se*, it is critical for your AIMS compliance and long-term performance and credibility to set up a framework that helps you address the issue of LLM audit. If you are developing, or using applications built on the top of LLM, then you might be part of the larger ecosystem of LLM and therefore it is critical to address issues of compliance.

A good framework for LLM audit is the framework suggested by authors Mökander, J., Schuett, J., Kirk, H. R., & Floridi (2023). This framework provides a three-layer approach to LLM audit.

1. **The audit of the governance structure of the technology provider**. According to the authors, the audit framework should focus first on the technology provider's governance framework. In this layer of audit, the assessment addresses issues of the process by which LLM gets produced, and the quality by which technology providers

create foundation models. In this stage, you look specifically at the following questions:

- How does the organization manage the production of LLM?
- What management systems are involved in the LLM production and deployment?
- What incentive procedures are in place?
- How does the technology provider account for issues of risk?
- What is the overall quality management process of LLM development?
- How does the LLM provider document the process of LLM capabilities?

2. **Model audit**. The second pillar of the LLM auditing process, according to Mökander et al. (2023), is the audit of the model itself. Here the audit looks at the following issues.
   - The limitation of the model: What are some of the limitations that the model can and cannot do? This is a crucial question for developers of API, so they are aware of how the model functions.
   - Information about the capacity of the model: What is the capability of the model? This is also a critical question for companies building applications on top of LLM as it helps them solve design issues.
   - Performance and robustness of the model: How does the model behave under different prompts and circumstances?
   - Security and vulnerability issues of the model: How does the model handle adversarial attacks and to what extent the data used to train the model is protected?
3. **Application Audit**. A third component of the audit framework suggested by Mökander et al. (2023) is to audit the application built on top of the model itself. The audit of the application should focus on compliance with the regulatory environment and the ethical framework of the developer's company. Applications audit, according to Mökander et al. (2023), should address two critical questions:
   - Is the intended use of the application legal, ethical, and subscribe to the ethical principles of the large language model's provider?
   - What is the impact of the application on users, communities, and society?

Finally, it is important to mention that the most critical contribution of this framework is that it addresses the critical issue of how and in what way LLM should be audited. The framework also provides a comprehensive approach that links the entire supply chain of large language model design and deployment: Technology providers and application developers.

Set Up an Internal Audit Program

The second step in the performance evaluation of the AIMS required by ISO/IEC 42001 is to set up an internal audit program. The purpose of the audit program is to check if the AIMS complies with the requirements of the organization and the ISO/IEC standard 42001. For the AIMS audit to comply with the standard, it has to achieve three goals:

- Achieve conformity of the AIMS with the organizational AI requirements.
- Meet the requirements set by the current standard ISO/IEC 42001.
- The AIMS is effectively implemented and maintained.

For audit programs to work effectively they need to take into consideration the organization's context. Factors such as business environment, legal context, size, the economic sector, and previous audits all need to be accounted for. In large organizations with a history of IT legacy, for instance, the audit program may need to be more thorough. The goal is to identify and spell out the practices that may impede the effective functioning of the AIMS and may create more risks for the organization.

There are many internal audit programs that you can use, in addition to the framework suggested by the ISO/IEC 42001, depending on the context of your organization and how it fits that specific framework.

A recent framework that received a lot of attention lately is the AI Risk Management Framework published by the National Institute of Standards and Technology in 2023 (NIST AI RFM 1.0). This framework is a comprehensive one and addresses many relevant aspects, such as effectiveness, trustworthiness, and responsible AI criteria. NIST AI RFM provides tools that can add practical value to the auditing process. The NIST RMF is made of four major steps: Govern, Map, Manage, and Measure. Let's take a look at these steps.

- **Govern step**: In this step, the audit needs to focus on how the audit program addresses the regulatory environment of the AIMS. This is a challenge for the auditors as the legal requirements and regulations are shifting constantly. In this step, auditors need to take into consideration specific new and emerging regulations that address AI specifically, but also exciting laws and regulations that address health and safety, discrimination, and consumer protection.

- **Mapping step**: In the mapping step, the goal of the auditor is to spell out and identify the links that connect to different parts of the AIMS. In other words, here you map out how and in what way your AIMS affects the different components that it interacts with internally and externally. This is a challenging step. As a general-purpose technology, AIMS interacts with many parts of the organization and is impacted by many parts of the organization. So, an effort must be made to identify how the AIMS gets affected (by the organization context, the culture, and the vision of the leadership) and how, in turn, the AIMS impacts the context and the wider society.
- **Measuring step**: According to the NIST RMF the measurement is a critical component of the audit program. Measuring the risks surrounding the AIMS, like errors and adverse impacts, helps the audit program build consistency and credibility. In this step, a focus on not only the competence and skills but also the objectivity and the independence of the audit program is critical to the success of the overall audit program.
- **Manage step**: In the last step, the organization needs to spell out how it intends to measure the risks identified. You can focus on the following actions:
  - How do the metrics identified fit the overall context of the organization and the audit process?
  - How does the organization intend to prioritize the risks and how it intends to treat them?
  - How do you document the way risks have been identified and treated?

Management Review

Once the audit is completed, management needs to review the results of the audit and act upon it. The goal of management review is to provide the audit program with a stamp of legitimacy. When management approves and signs off on an audit program, the program becomes an official program of the organization and a roadmap for future improvement. Management review is also critical. It shows how leadership engages with the AIMS. The leadership tone, guidance, and influence are exerted through the review of the AI program audit.

For an AIMS review to comply with ISO/IEC 42001, the review must include the following:

- The results of the previous review and audit. This is an important input that allows management to compare previous and current audits to identify gaps and improvements.
- Changes that have taken place since the last audit cycle and how the organization responded and adjusted to them. Changes that relate not only to laws and regulations but also to the economic and business environment must be spelled out so management can comprehend the improvements that are recommended. In this context, a must-do action is to address the regulation changes and how the new public interest and scrutiny are addressed by the AIMS.
- How stakeholders' needs and expectations have changed since the last time the audit was conducted, and in what way these changes have impacted the effectiveness of the AIMS? A good practice here is to map out all AIMS stakeholders and show how AIMS has impacted them since the last time the audit was completed.
- What opportunities for improvement can the organization address based on the measurement and the nonconformity issues the audit has identified? This would be your future improvement roadmap.

Finally, the entire audit results must be documented. This is a requirement of ISO/IEC42001:2023.

TOOLBOX 23

WHAT IS AN ALGORITHMIC AUDIT?

In recent years, a new concept emerged in the audit literature. The concept of "algorithmic audit" is a new concept that refers to the process of inspecting how algorithm models are built and trained and how they behave. Algorithms, which are a set of instructions built on a set of data, may contain errors that may affect their behavior and their integrity, and may make them behave in a biased and discriminatory way. So, checking inspecting and testing how the algorithm was built and the type of data it was trained on has emerged in recent years as a critical issue in the conversation about overall AI systems.

A study conducted by A. Hasan, S. Brown, J. Davidovic, B. Lange, and M. Regan (2022) on algorithmic bias and risk assessment provides a good framework for how to audit algorithms. First, the authors differentiate between two related concepts: Ethical risk assessment of AI systems in general, and a narrower more technical assessment that focuses on algorithmic bias assessment. While in the first type of assessment, the purpose is to look at the sociotechnical environment in which the AI is being deployed, in the algorithmic assessment, the goal is to inspect the inner functioning of the algorithm itself.

In the ethical assessment, you need to first identify key stakeholders and the potential harm that may impact them. In this step, the authors suggest that you start with broad research to identify the possible harm that the algorithm being studied may have on people. This is a generic identification that does not necessarily involve the algorithm being studied. The purpose of this is simply to be aware of the possible and probable harm that may occur.

In the second step, according to Hassan et al. (2022), you focus on two important elements: Data collection methods, how data was processed, managed, and labeled; and the testing metrics that were defined before the building of the ML model.

Chapter 11

Clause 10: Improvement

PURPOSE

The purpose of this chapter is to explain the AIMS continuous improvement process. The chapter links the AIMS and improvement process to the PDCA process, so improvements are standardized and continuous.

OVERVIEW

The chapter defines the AIMS process improvement. The reader will find an explanation of the required actions to implement the ISO clause to meet standard requirements. The chapter also explains a few concepts and terminologies that relate to continual improvement and provides a model to follow for AIMS improvement purposes. The reader will find an explanation of the PDCA process and how it applies to AIMS improvement. Links between improvement and performance evaluations are also identified.

BENEFITS

At the end of this chapter, you will be able to:

- Learn how to conduct an AIMS continual improvement process.
- Understand how to integrate improvement concepts into the design of the overall improvement process.

DOI: 10.4324/9781003463979-13

- Understand how to integrate the CAPA framework into the overall improvement process.
- Learn how to identify links between different stages of the PDCA model for a better and more sustainable AIMS improvement process.

Introduction

Section 10 of the standard ISO/IEC 42001 addresses the issues of improvement, deviation, nonconformity, and continual improvement of the AIMS. At the end of the performance evaluation, and as a result of the management review, you should have a relatively good idea about the weaknesses and the strengths of your AIMS.

To help you understand how to build a continual improvement plan to improve your AIMS and meet the standard requirement, I will adopt a PDCA approach to answer the following critical required questions:

- How do you identify and measure the improvement needed to enhance the strength of AIMS? (Plan and Do)
- How do you implement the improvement you identified to address the shortcomings of the AIMS? (Check)
- What lessons you have learned that may help you correct the deviations and the nonconformity of your AIMS? (Act)

But before we get into the details of the PDCA and the continual improvement plan, let us go over a few terminologies used in Section 10 of the standard ISO/IEC 42001. The section uses two terminologies: Deviation and nonconformity.

- **Deviation** is an undesirable result or a gap between the actual performance and the initial target performance of the system. For your AIMS, many deviations may happen during the AI project life cycle, so it is critical to have gating questions and KPIs at the end of each stage of the AI life cycle.
- **Nonconformity**, on the other hand, is simply the nonfulfillment of a required action such as not having a documented AI policy, or not

being able to define and document the roles and responsibilities of different actors engaged with the AIMS.

When there is a deviation or a nonconformity, there are three types of actions you need to take according to the standard.

- Simply accept the consequences of the nonconformity and learn from it while at the same time taking action to correct it.
- Understand why the deviation happened in the first place by examining and analyzing the circumstances that produced it.
- Update or make changes to the AIMS based on your understanding of the shortcoming.

Of these three suggestions, the second one is the most important recommendation as it helps you conduct a root-cause analysis to understand the source of the deviation, where it comes from during the AI project life cycle, and how and why the organization missed it in the first place. It also helps you identify future actions that you may need to take to correct the shortcomings.

Corrective Actions and Preventive Actions Framework

Corrective Actions and Preventive Actions (CAPA) is a framework that helps managers structure efforts to conduct a root-cause analysis of the deviation and take actions to understand what could trigger the deviation or nonconformity in the future. Professor Nathan Conover (2020), from the consulting firm Pathways, defines CAPA as the "immune system of the organization." CAPA helps you identify the risk before it happens and disseminate a culture of prevention among team members.

Corrective Actions

Corrective actions help you identify deviation and nonconformity in the AIMS in the following ways.

- When there is a deviation, CAPA allows you to initiate corrective actions, which should allow you to take immediate actions to eliminate

the problem. For instance, if the data quality does not meet your definition and has been identified as a deviation, then you need to reconsider the source of your data.
- Corrective actions help you eliminate the cause of the deviation or the cause of the nonconformity permanently. In the root-cause analysis, you should have gathered data on what happened, how it happened, why it happened, and what should be done to avoid the problem in the future. Generally speaking, the root-cause analysis addresses three types of problems: human error (an employee not following the protocol), a technical issue (how the model is being trained), or a process error (how data is wrangled).

Preventive Actions

Once you correct immediately the deviation in the preventive actions, you are putting in place a system that helps you predict when a problem is going to happen, and how you could prevent it. It is like putting in place a trigger and an alarm system that sends signals when a problem is about to occur. Of critical importance to CAPA methodology and the improvement system in general is the documentation. Documenting the source of the problem, how it was identified, and how it was solved is critical to preventing the same problem from occurring in the future.

Continual Improvement: The PDCA Approach

One of the most known frameworks in the improvement projects field is the Deming Cycle of Improvement: Plan-Do-Check-Act. The PDCA is a highly interactive framework. It treats improvement, not as one event, but, rather, as a continuous effort that spans over the entire lifecycle of the AI project. PDCA is also a great tool because it helps link different teams engaged with the AI project in one framework. The PDCA framework is a flexible and simple framework that you can apply to many processes of the business.

Below I will go over these five PDCA steps:

1. **Plan**: The first step of the PDCA cycle is planning. The planning of the improvement project requires many actions, the chief of which is to

define the business problem and to get an agreement from all team members engaged in the AI project on what constitutes a problem. This is not an easy task as a typical AI project involves many moving parts, with each part being a mini project by itself. So, getting people to agree on the improvement plan is critical. Following are some of the questions that you need to address in this step.

- **Define the project business goal**: In this stage, you need to have both technical teams and business teams define the business problem. A problem described by a team of data scientists may be viewed differently by a team working on customer experience. The goal is to get an agreement from everybody on the pain points and the issues the AI project is expected to address.
- **Clarify the objective of the AI improvement project**: In this stage, it is critical that you get an agreement on the objectives of the improvement project you intend to implement. Listing all the AI project improvement requirements, goals, and objectives would help you create a roadmap for different teams involved in the project improvement.
- **Create metrics and milestones**: It is important that you define the metrics you will use to assess the AI project improvement and how these metrics will be met so everyone on the team understands the deliverables.
- **Type of resources available**: Identify the resources needed for the AI project. Think of financial resources, like the budget allocated to the AI project, physical resources, like computing power and the networking infrastructure, and human resources, the engineers, and data scientists that the project will need.

2. **Do**: The second step of the PDCA model is the implementation. Here you are implementing the different actions you planned, such as:
   - Data collection: How data is collected, processed, and managed overall.
   - How is the model working vs. how it is supposed to work
   - Risks you have identified and how far you were in your prediction of risks.
   - Test and experiment to make sure that the ML project is achieving the results planned.
3. **Check**: In the third step of the PDCA model, the goal is to verify and check that your AIMS is working as it is intended to. This is the step in which you will be testing and monitoring the AIMS processes. You can

proceed in two steps. First check the governance of the AIMS, or the sociotechnical requirements, such as leadership, the policy, the risk management, and how the overall AIMS is performing. And, in the second step, focus on the behavior of the algorithm, and whether the technical side of the AIMS is producing the results hoped for. You need to make sure that your AI algorithm is meeting the testing and measurement requirements that you have identified at the beginning of the PDCA cycle.

Following are sample questions that you may ask when testing the sociotechnical dimension of your AIMS:

- Do we have an AI policy, and is it updated and documented?
- Are the measurements we set to monitor the alignment and the coherence of the AIMS working as they were intended to, and what KPIs have been used to measure that?
- Are the metrics we set to assess the impact reliable and yielding results we can rely on?

Then you move to the testing of the algorithm behavior itself. This is the technical analysis of your AIMS. Ask the following questions:

- Is the trained AI model performance working according to the metrics we have identified?
- Can we validate the model's accuracy against other relevant metrics to make sure that the algorithm is working as it is supposed to?
- What issues the testing has identified that we can act upon?

4. **Act**: The act step in the PDCA framework refers to the feedback and improvement actions that you plan to take to correct nonconformities and improve the system. Depending on the outcome of the previous step, here you are working on improving the AIMS based on the results of the checking and monitoring you did previously. Here also you can proceed in two ways. Focus on the governance issues, then tackle the technical aspect of data and algorithms. Figure 11.1 provides a PDCA approach to improving the AI project.

Conclusion

The age of AI has ushered in a new era that requires a paradigm shift in the way we lead and manage change. The acceleration of change requires that we learn, adapt, adjust, and develop new mental frameworks faster than we used to. As a learning tool, standards help spread the word and establish

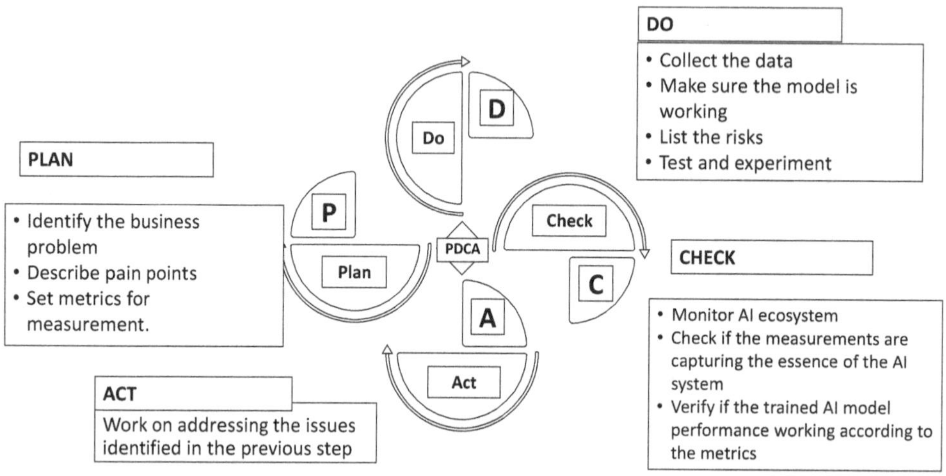

Figure 11.1 Continual Improvement: The PDCA Approach to Improving the AI Project Life Cycle.

the foundation of good practices to mature and reproduce success at scale. This is what I believe ISO/IEC 42001 will do to the AI systems, and this is why I think organizations should adopt 42001 frameworks as fast as possible, so they can build the foundations of a dynamically shifting technology.

In this book, I wanted to emphasize four key ideas.

First, the transformation we are going through, and while it seems to be led by technology, it is affecting all facets of the economy. Organizations that are still stuck in the old paradigm of managing the technology-led change as an IT issue will struggle and eventually fall behind. The type of transformation we live in today, although it's called digital transformation, requires a comprehensive approach that moves many things at the same time. From setting up a business strategy to creating a data-quality framework to managing AI projects, talent mobility, and employee growth. The faster, holistic, and agile approach is what is needed for the age of AI.

Second, and in discussing the different clauses of the standard, my goal was to show you how to implement the standard and put together the foundation of a viable AI management system that can respond to the changing environment in which it lives. While the standard clauses are straightforward and can be understood by most compliance officers and quality managers familiar with ISO standardization procedures, showing how to implement the clauses, how to create the documentation needed, and how to streamline the system's processes and procedures required by the standard is one of the added values of this book.

Third, a key idea of the standard is around AI risk. This is why you will find the terms that relate to risk, such as *risk assessment*, *risk treatment*, *impact assessment*, and *overall risk management*, more often cited in the standard. I have devoted to these concepts an entire chapter, in addition to explaining them when addressing AI technology and generative AI. My goal was to help you understand the intricacies of risk and how it connects with other enterprise risks.

Finally, and to give the reader a more comprehensive view of how this standard can promote an emerging technology and mature its practices, I went beyond ISO IC 42001. While the bulk of the book is about AI management systems, as explained by ISO/IEC 42001, I also addressed other standards that form the overall ecosystem of AI technology and that have been jointly developed by ISO and IEC.

Appendix

50 Most Important Terms in AI and ISO Standards

Following is a list of 50 words that teams involved with ISO/IEC 42001 need to know. It is a combination of technical words that relate to computer science and AI, quality terminology and ISO language, and some cognitive bias terms that have become critical to know for data scientists and engineers involved in the production of AI models.

1. **A group attribution error**: The error of believing that the attribute of an individual group member reflects the entire group. This can affect the quality of the sampling.
2. **Accreditation body**: An organization or an agency that has the authority to certify certain processes or capabilities.
3. **Accreditation**: A recognition, through a certification process, by a recognized body, of certain capabilities.
4. **Algorithm**: A set of instructions a computer follows to execute tasks.
5. **Anchoring bias**: A bias in which you focus on one piece of data using it as a reference for decisions without proof that the data is valid or accurate.
6. **Artificial Intelligence**: A generic concept that describes the ability of machines to perform human-like tasks.
7. **Autonomous systems**: Systems that can act on their own without human intervention.
8. **Baseline measurement**: A reference assessment that helps set the goal for the future improvement required.

9. **Behavioral analytics**: An emerging field that studies people's behavior on digital platforms.
10. **Biometrics**: The use of people's physical traits to identify them, like facial recognition.
11. **Business Process Reengineering (BPR)**: The reconfiguration of the business process for improvement purposes.
12. **Causation**: The direct relationship between two variables, like a change in variable A directly causes a change in variable B.
13. **Certification**: The process of recognizing a capability by an accredited authority.
14. **Chatbots**: A program that mimics human-to-human communication and is set to communicate with people.
15. **Confirmation bias**: A cognitive bias in which you seek information to confirm a pre-existing belief.
16. **Conformance**: An affirmation that a process, service, or product has met the required specifications.
17. **Conformity assessment**: An assessment process to attest that a product, service, or process has met the requirements.
18. **Correlation coefficient**: The level of interdependence between two variables.
19. **Data aggregation**: The process of collecting data from different sources to use it for learning and decision-making purposes.
20. **Data science**: An emerging interdisciplinary field that is at the intersection of statistics and computer sciences. It uses statistical concepts to solve problems and program computers.
21. **Deep learning**: A subset of AI specialization that mimics brain architecture to make decisions.
22. **DMAIC**: Define, measure, analyze, improve, and control is a Six Sigma methodology used in continuous improvement.
23. **Effectiveness**: The ability to achieve better results with better impact.
24. **Efficacy**: The ability to achieve the results planned for.
25. **Efficiency**: The ability to achieve results with optimal resources.
26. **Explainability**: A term associated with the ability to explain an AI model, so it makes sense to the average person (different from interpretability).
27. **False causality**: Mistakenly linking cause and effect.
28. **Governance**: The set of rules and procedures used by an organization to govern.
29. **Illusion of validity**: An error in which you believe the accuracy of your judgment when information seems to be plausible.

30. **Illusory correlation**: The error of believing that two unrelated variables are connected. This can be an issue when building models using low-quality data.
31. **Impact assessment**: The process of assessing and identifying the impact of a technology on a specific category of users. Impact assessment is different from risk assessment.
32. **Implicit bias**: Called also subconscious bias, it is an error in which a course of action is subjectively chosen without any objective data that corroborates the choice.
33. **In-group bias**: Called also similar to me bias, is an error in which people similar to the researchers are favored. This also can impact data selection.
34. **Interpretability**: The ability of a system to follow a cause–effect logic in its working. In other words, you can link what affected B to the causes produced by A in a meaningful way.
35. **Machine learning**: A subset of AI in which machines follow instructions and learn from the instructions to accomplish tasks and develop new capabilities.
36. **Metadata**: Information and data that describe other data (data about data).
37. **Mind mapping**: A methodology that allows for ideas to be connected to produce a map that helps people externalize and visualize connections between concepts and events.
38. **Natural Language Process (NLP)**: The discipline in computer science that focuses on creating human life conversations by machines.
39. **Neural network**: A network of nodes that mimic the human brain in its decision-making process.
40. **Outcome bias**: The tendency to judge an action by the result rather than looking at the quality of the process by which the outcome was achieved.
41. **Overfitting**: The statistical error in which you tend to force an explanation into a dataset.
42. **Plan-Do-Check-Act (PDC A)**: A structured approach to conducting improvement projects.
43. **Process Map**: A flow chart that breaks down the different steps of a process.
44. **Reinforcement learning**: A machine learning method that teaches machines how to make decisions by rewarding desired behaviors and punishing undesired ones.

45. **Risk management**: The process by which risks are identified, categorized, and mitigated.
46. **Risk mitigation**: The process of identifying ways to reduce the impact of risk.
47. **Sampling error**: An error in which the sample is not representative of the target population.
48. **SIPOC diagram**: Supplier, input, process, output, and customer is a Six Sigma method that helps identify the contribution of each component to the process.
49. **Six Sigma**: A quality methodology designed to reduce variability in a process.
50. **Validity**: The measurement of what was intended to be measured.

Bibliography

Abrahamson, E. (2000). Change Without Pain. Harvard Business Review. July–August 2000.

Accenture (2018). Process reimagined. Together, people and AI are reinventing business processes from the ground up. Retrieved from https://www.accenture.com/us-en

Anderson, J., & Rainie, L. (2023). *As AI Spreads, Experts Predict the Best and Worst Changes in Digital Life by 2035*. Pew Research Center. Retrieved from https://www.pewresearch.org/

BCG (2022). Responsible AI for an Era of Tighter Regulations. Retrieved from https://www.bcg.com/publications/2021/the-four-stages-of-responsible-ai-maturity

Bowers, G. (2023). The Age of Ai. Franklin Templeton. Retrieved from. https://www.franklintempleton.com/articles/2023/disruption/age-of-ai

Broadband Commission (2022). Artificial Intelligence and Digital Transformation: Competencies for Civil Servants. Retrieved from https://link.springer.com/article/10.1007/s11846-022-00613

CFPB (2023). Joint Statement on Enforcement Efforts Against Discrimination And Bias In Automated Systems. Retrieved from https://www.consumerfinance.gov/about-us/newsroom/cfpb-federal-partners-confirm-automated-systems-advanced-technology-not-an-excuse-for-lawbreaking-behavior/

Cohen, R., & Cohen, R. (2017). *Highly Stratified Occupations and the Digital Economy*. Economic Strategy Institute.

Conover, C. (2020). What Is Corrective and Preventive Action? Retrieved from https://www.smartsheet.com/corrective-and-preventive-action

Consulting US (2021). How digital transformation leaders are standing out. Retrieved from https://www.consulting.us/news/4380/how-digital-transformation-leaders-are-standing-out.consumer harms of generative AI. Retrieved from www.forbrukerradet.no/ai

Davenport, T. H., & Ronanki, R. (2018). Artificial intelligence for the real world. *Harvard Business Review*, 96(1), 108–116.

De Silva, D. and Alahakoon, D. (2021). *An Artificial Intelligence Life Cycle: From Conception to Production*. Centre for Data Analytics and Cognition (CDAC). La

Trobe University, Victoria, Australia. Retrieved from https://www.researchgate.net/publication/354268391_An_Artificial_Intelligence_Life_Cycle_From_Conception_to_Production

Deloitte. (2018). Artificial Intelligence Delivers, but Missteps Can Yield 'Bridges to Nowhere.' Retrieved from https://www2.deloitte.com/us/en/pages/about-deloitte/articles/press-releases/deloitte-launches-2018-artificial-intelligence-in-the-enterprise-report.html

Deloitte University Press. Rewriting the Rules for the Digital Age: 2017 Deloitte Global Human Capital Trends. Retrieved from https://www2.deloitte.com/content/dam/Deloitte/global/Documents/HumanCapital/hc-2017-global-human-capital-trends-gx.pdf

Deloitte. (2022). Fueling the AI transformation: Four key actions powering wide-spread value from AI, right now. Retrieved from https://www2.deloitte.com/us/en/pages/consulting/articles/state-of-ai-2022.html

Demircan, M. (2023). Deployers of High-Risk AI Systems: What Will Be Your Obligations Under the EU AI Act? LSTS Research Group, Vrije Universiteit Brussel. Retrieved from https://competitionlawblog.kluwercompetitionlaw.com/2023/06/02/deployers-of-high-risk-ai-systems-what-will-be-your-obligations-under-the-eu-ai-act/

DeSmith, A. (2015). The Keys to Organizational Agility. Retrieved from https://www.mckinsey.com/business-functions/organization/our-insights/the-keys-to-organizational-agility

Di Vinadio, T., Noordt, C., Castillo, C., Avila, R. (2022). Working Group Report on AI Capacity Building Artificial Intelligence and Digital Transformation Competencies for Civil Servants. Broadband Commission on Sustainable Development. Retrieved from https://www.broadbandcommission.org/publication/artificial-intelligence-and-digital-transformation/

Diamandis, P., & Kotler, S. (2012). *Abundance: The Future Is Better Than You Think*. Free Press.

Digital Curation Centre (2020). The Role of Data in AI Report for the Data Governance Working Group of the Global Partnership of AI. Retrieved from https://www.google.com/url?sa=t&rct=j&q=&esrc=s&source=web&cd=&ved=2ahUKEwjfroi7mLD-AhU2VaQEHUQ7DfAQFnoECB0QAQ&url=https%3A%2F%2Fgpai.ai%2Fprojects%2Fdata-governance%2Frole-of-data-in-ai.pdf&usg=AOvVaw1LlgZH05DbNw5V41T8xBif

Dotan, R. (2023). Regulation of AI In the United States: A Repository. Retrieved from https://www.ravitdotan.com/copy-of-ai-due-diligence

Edwards, L. (2022). Expert explainer. The EU AI Act: A Summary of its Significance And Scope. Retrieved from https://www.adalovelaceinstitute.org/resource/eu-ai-act-explainer/

EEOC (2023). Select Issues: Assessing Adverse Impact in Software, Algorithms, and Artificial Intelligence Used in Employment Selection Procedures Under Title VII of the Civil Rights Act of 1964. Retrieved from https://www.eeoc.gov/select-issues-assessing-adverse-impact-software-algorithms-and-artificial-intelligence-used

Fade, P. (2020). *Customer Centricity: Focus on the Right Customers for Strategic Advantage*. Wharton School Press.

Fagella, D. (2020). AI Report. Retrieved from https://emerj.com/dan-faggella/

GovLoop. (2022). *How to Reduce the Complexity of Data Initiatives*. A Market Trends Report. Retrieved from https://go.govloop.com/complex-data-confirm.html?aliId=eyJpIjoiQXhiS0JORFE3NmJXeStDNCIsInQiOiJpSkVmTzFFT01obTl3WWZaMHJ0UXBRPT0ifQ%253D%253D

GSA (2023). AI Guide for Government: A living and evolving guide to the application of Artificial Intelligence for the U.S. federal government. Retrieved from https://coe.gsa.gov/coe/ai-guide-for-government/print-all/index.html

Hao, K. (2019). This is how AI bias really happens—and why it's so hard to fix. Retrieved from https://www.technologyreview.com/2019/02/04/137602/this-is-how-ai-bias-really-happensand-why-its-so-hard-to-fix/

Hasan, A., Brown, S., Davidovic, J., Lange, B., & Regan, M. (2022). Algorithmic Bias and Risk Assessments: Lessons from Practice. Retrieved from Digital Society (2022) https://doi.org/10.1007/s44206-022-00017-z

Haugeland, J. (1985). *Artificial Intelligence: The Very Idea*. MIT Press.

Henderson, P., Li, X., Jurafsky, D., Hashimoto, T., Lemley, M. A., & Liang, P. (2023). *Foundation Models and Fair Use*. Stanford University. Retrieved from https://arxiv.org/abs/2303.15715

High-Level Expert Group on Artificial Intelligence. (2020). Ethics Guidelines For Trustworthy AI. Retrieved from https://digital-strategy.ec.europa.eu/en/policies/expert-group-ai

IBM Analytics (2020). The Data Science Skills Competency Model: A blueprint for the growing data scientist profession. Retrieved from https://www.ibm.com/analytics

IDC (2023). 2023 Insight Intelligent Technology Report: Are We Entering a New Era of Innovation? Retrieved from https://www.insight.com/en_US/content-and-resources/gated/insight-intelligent-technology-report-ac1252.html

Institute and Faculty of Actuaries and Royal Statistical Society. (2019). A Guide for Ethical Data Science A collaboration between the Royal Statistical Society (RSS) and the Institute and Faculty of Actuaries (IFoA). Retrieved from https://actuaries.org.uk/standards/data-science-ethics/

Institute of Director. (2022). AI in the Boardroom: The essential questions for your next board meeting. Retrieved from https://www.iod.com/resources/blog/science-innovation-and-tech/ai-in-the-boardroom-the-essential-questions-for-your-next-board-meetin/

ISO/IEC TS 4213:2022 Information technology — Artificial intelligence — Assessment of machine learning classification performance.

ISO/IEC CD 5259-1 Artificial intelligence — Data quality for analytics and machine learning (ML) — Part 1: Overview, terminology, and examples.

ISO/IEC CD 5259-2 Artificial intelligence — Data quality for analytics and machine learning (ML) — Part 2: Data quality measures.

ISO/IEC DIS 5338 Information technology — Artificial intelligence — AI system life cycle processes.

ISO/IEC DIS 5339 Information technology — Artificial intelligence — Guidance for AI applications.
ISO/IEC AWI TS 6254 Information technology — Artificial intelligence — Objectives and approaches for explainability of ML models and AI systems.
ISO/IEC FDIS 8183 Information technology — Artificial intelligence — Data life cycle framework.
ISO/IEC 20546:2019 Information technology — Big data — Overview and vocabulary.
ISO/IEC 22989:2022 Information technology — Artificial intelligence — Artificial intelligence concepts and terminology.
ISO/IEC 23894:2023 Information technology — Artificial intelligence — Guidance on risk management.
ISO/IEC TR 24028:2020 Information technology — Artificial intelligence — Overview of trustworthiness in artificial intelligence.
Information technology — Artificial intelligence — Overview of ethical and societal concerns.
ISO/IEC 38507:2022 Information technology — Governance of IT — Governance implications of the use of artificial intelligence by organizations.
ISO/IEC DIS 42001 Information technology — Artificial intelligence — Management system.
ISO/IEC AWI 42005 Information technology — Artificial intelligence — AI system impact assessment.
ISO/IEC CD 42006 Information technology — Artificial intelligence — Requirements for bodies providing audit and certification of artificial intelligence management system.
Kazim, E., Mendes Thame Denny, D., & Koshiyama, A. (2021). AI auditing and impact assessment: according to the UK information commissioner's office. Retrieved from https://link.springer.com/article/10.1007/s43681-021-00039-2
Kurzweil, R. (1990). *The Age of Intelligent Machines*. MIT Press.
Lee, M. Seng, A. Floridi, L., & Singh, J. (2020). Formalizing trade-offs beyond algorithmic fairness: lessons from ethical philosophy and welfare economics (July 31, 2020). Available at SSRN: https://ssrn.com/abstract=3679975 or http://dx.doi.org/10.2139/ssrn.3679975
Leswing, K. (2023). *ChatGPT and generative AI are booming, but the costs can be extraordinary*. CNBC. Retrieved from https://www.cnbc.com/2023/03/13/chatgpt-and-generative-ai-are-booming-but-at-a-very-expensive-price.html
Mahowald, K., Ivanova, A. A., Blank, I. A., Kanwisher, N., Tenenbaum, J. B., & Fedorenko, E. (2023). Dissociating language and thought in large language models: A cognitive perspective. arXiv preprint arXiv:2109.06129. Retrieved from https://www.researchgate.net/publication/367217861_Dissociating_language_and_thought_in_large_language_models_a_cognitive_perspective
Maslej, N., Fattorini, L., Brynjolfsson, E., Etchemendy, J., Ligett, K., Lyons, T., Manyika, J., Ngo, H., Niebles, J. C., Parli, V., Shoham, Y., Wald, R., Clark, J., & Perrault, R. (2023). The AI Index 2023 Annual Report. AI Index Steering

Committee, Institute for Human-Centered AI, Stanford University, Stanford, CA, April 2023. Retrieved from https://creativecommons.org/licenses/by-nd/4.0/?ref=chooser-v1

McCarthy, J. (1958). Recursive Functions of Symbolic Expressions and Their Computation by Machine, *Part I. Communications of the ACM*, *3*(4), 184–195.

McChrystal, S., Collins, T., Silverman, D., & Fussell, C. (2015). *Team of Teams: New Rules of Engagement for a Complex World.*

McKinsey & Company (2018). Unlocking success in digital transformations. Retrieved from https://www.mckinsey.com/capabilities/people-and-organizational-performance/our-insights/unlocking-success-in-digital-transformations

McKinsey & Company (2020). Enterprise agility: Buzz or business impact? Retrieved from https://www.mckinsey.com/capabilities/people-and-organizational-performance/our-insights/enterprise-agility-buzz-or-business-impact

McKinsey & Company (2023a). The economic potential of generative AI: The next productivity frontier. Retrieved from https://www.mckinsey.com/capabilities/mckinsey-digital/our-insights/the-economic-potential-of-generative-ai-the-next-productivity-frontier#key-insights

McKinsey & Company (2023b). State of the AI Report 2022. Retrieved from https://www.mckinsey.com/capabilities/quantumblack/our-insights/the-state-of-ai-in-2022-and-a-half-decade-in-review

Merali, S., Merali, A., & Hawksbee, A. (2023). The-Generative AI Revolution: Opportunities, Shocks, and Risks. Retrieved from www.ukonward.com

Mökander, J., Schuett, J., Kirk, H. R., & Floridi, L. (2023). Auditing Large Language Models: A Three-Layered Approach. Retrieved from https://ssrn.com/abstract=4361607

Müller, C. (2022). World Robotics 2022 Industrial Robots, IFR Statistical Department, VDMA Services GmbH, Frankfurt am Main, Germany, 2022. Retrieved from https://ifr.org/

Murphy, J. (2023). How businesses can measure AI success with KPIs. TechTarget. Retrieved from https://www.techtarget.com/searchenterpriseai/tip/How-businesses-can-measure-AI-success-with-KPIs

Norwegian Consumer Council. (2023). Ghost in the machine: Addressing the Addressing the consumer harms of generative AI. Retrieved from www.forbrukerradet.no/ai

Petrova, S. (2022). Game-Changing Digital Transformation Frameworks. Retrieved from https://adevait.com/blog/workplace/game-changing-digital-transformation-frameworks

Phillips, J., Hahn, C., Fontana, P., Yates. A., Greene, K., Broniatowski, D., & Przybocki, M. (2021). *Four Principles of Explainable Artificial Intelligence.* National Institute of Standards and Technology. Retrieved from https://doi.org/10.6028/NIST.IR.8312

PWC (2017). Sizing the prize: PwC's Global Artificial Intelligence Study: Exploiting the AI Revolution. What's the real value of AI for your business and how can you capitalize? Retrieved from https://www.pwc.com/gx/en/issues/data-and-analytics/publications/artificial-intelligence-study.html

Ransbotham, S., Candelon, F., Kiron, D., LaFountain, B., & Khodabandeh, S. (2021). The Cultural Benefits of Artificial Intelligence in the Enterprise. *MIT Sloan Management Review* and Boston Consulting Group.

Roslansky, R. (2021). Navigating the Great Reshuffle. Retrieved from https://www.linkedin.com/pulse/navigating-great-reshuffle-ryan-roslansky/

Schuett, J., Dreksler, N., Anderljung, M., McCaffary, D., Heim Emma, L., & Ben Garfinkel, B. (2023a). Towards best practices in AGI safety and governance: A survey of expert opinion. Retrieved from https://arxiv.org/abs/2305.07153

Schuett, J., Dreksler, N., Anderljung, M., McCaffary, D., Heim, L., Bluemke, E., & Garfinkel, B. (2023b). Towards best practices in AGI safety and governance: A survey of expert opinion. Centre for the Governance of AI. arXiv:2305.07153v1 [cs.CY] 11 May 2023. Retrieved from https://arxiv.org/abs/2305.07153

Smuha, N. A. (2021). Beyond the individual: governing AI's societal harm. *Internet Policy Review*, 10(3). Retrieved from https://doi.org/10.14763/2021.3.157

Soendergaard, P. (2020). AI Governance – What Are the KPIs? And Who is Accountable? Retrieved from https://2021.ai/ai-governance-kpi/

Stobierski, T. (2021). Data Wrangling: What It Is & Why It's Important. Retrieved from https://online.hbs.edu/blog/post/data-wrangling

Storrar, T. (2021). What are the infrastructure requirements for artificial intelligence? Retrieved from https://www.datacenterdynamics.com/en/opinions/what-are-the-infrastructure-requirements-for-artificial-intelligence/

The Center for Equity, Gender, and Leadership. (2020). Mitigating Bias in Artificial Intelligence: An Equity Fluent Leadership Playbook. Retrieved from https://haas.berkeley.edu/equity/

The Economist Impact (2023). Risk Reset: Shifting Focus From Reaction To Anticipation. Retrieved from https://impact.economist.com/new-globalisation/risk-reset/

The Economist Intelligence Unit. (2020). Staying ahead of the curve: The business case for responsible AI. Retrieved from https://www.eiu.com/n/staying-ahead-of-the-curve-the-business-case-for-responsible-ai/

Thompson, S. (2022). *Managing Machine Learning Projects From design to deployment Version 8*. Manning Publications Co. Retrieved from https://www.techtarget.com/searchenterpriseai/feature/Defining-requirements-key-to-manage-machine-learning-projects

Trianz. (2021). State of Digital Transformations Worldwide 2020. Retrieved from https://www.trianz.com/dt-2020-report

Vaswani, A., Shazeer, N., Parmar, N., Uszkoreit, J., Jones, L., Gomez, A. N., Kaiser, Ł., & Polosukhin, I. (2017). Attention is all you need. Retrieved from https://blogs.nvidia.com/blog/2022/03/25/what-is-a-transformer-model/

WEF (2019). Harnessing Artificial Intelligence for the Earth. Retrieved from https://www.weforum.org/press/2018/09/machines-will-do-more-tasks-than-humans-by-2025-but-robot-revolution-will-still-create-58-million-net-new-jobs-in-next-five-years/

Wheeler, T. (2023). The three challenges of AI regulation. The Brookings Institution.Retrievedfromhttps://www.brookings.edu/articles/the-three-challenges-of-ai-regulation/

World Economic Forum (2022). Empowering AI Leadership: AI C-Suite Toolkit. Retrieved from https://www.weforum.org/reports/empowering-ai-leadership-ai-c-suite-toolkit/?DAG=3&gclid=Cj0KCQjw27mhBhC9ARIsAIFsETG0TzmksO5j6woRfyOpB20j9fGNb9-raii6Nt-f94L3nP4NLHo7TFgaAksjEALw_wcB

World Economic Forum (2023a). Adopting AI Responsibly: Guidelines for Procurement of AI Solutions by the Private Sector. Insight report. Retrieved from https://www.weforum.org/reports/empowering-ai-leadership-ai-c-suite-toolkit/?DAG=3&gclid=Cj0KCQjw27mhBhC9ARIsAIFsETG0TzmksO5j6woRfyOpB20j9fGNb9-raii6Nt-f94L3nP4NLHo7TFgaAksjEALw_wc

World Economic Forum (2023b). The Presidio Recommendations for Responsible Generative AI. Retrieved from https://www.weforum.org/whitepapers/the-presidio-recommendations-on-responsible-generative-ai

Zeitgeist (2023). AI Readiness Report. Retrieved from https://scale.com/ai-readiness-report-2022

Index

Page numbers in *italic* refer to figures and page numbers in **bold** refer to tables.

N

O

P

R